DRIVER BEHAVIOUR AND ACCIDENT RESEARCH METHODOLOG

Human Factors in Road and Rail Transport

Series Editors

Dr Lisa Dorn
*Director of the Driving Research Group, Department of Human Factors,
Cranfield University*

Dr Gerald Matthews
Professor of Psychology at the University of Cincinnati

Dr Ian Glendon
*Associate Professor of Psychology at Griffith University, Queensland,
and is president of the Division of Traffic and Transportation Psychology
of the International Association of Applied Psychology*

Today's society must confront major land transport problems. The human and financial costs of vehicle accidents are increasing, with road traffic accidents predicted to become the third largest cause of death and injury across the world by 2020. Several social trends pose threats to safety, including increasing car ownership and traffic congestion, the increased complexity of the human-vehicle interface, the ageing of populations in the developed world, and a possible influx of young vehicle operators in the developing world.

Ashgate's 'Human Factors in Road and Rail Transport' series aims to make a timely contribution to these issues by focusing on the driver as a contributing causal agent in road and rail accidents. The series seeks to reflect the increasing demand for safe, efficient and economical land-based transport by reporting on the state-of-theart science that may be applied to reduce vehicle collisions, improve the usability of vehicles and enhance the operator's wellbeing and satisfaction. It will do so by disseminating new theoretical and empirical research from specialists in the behavioural and allied disciplines, including traffic psychology, human factors and ergonomics.

The series captures topics such as driver behaviour, driver training, in-vehicle technology, driver health and driver assessment. Specially commissioned works from internationally recognised experts in the field will provide authoritative accounts of the leading approaches to this significant real-world problem.

Driver Behaviour and Accident Research Methodology

Unresolved Problems

ANDERS AF WÅHLBERG
Uppsala University, Sweden

CRC Press
Taylor & Francis Group
Boca Raton London New York

CRC Press is an imprint of the
Taylor & Francis Group, an **informa** business

CRC Press
Taylor & Francis Group
6000 Broken Sound Parkway NW, Suite 300
Boca Raton, FL 33487-2742

First issued in paperback 2017

© 2009 by Anders af Wåhlberg
CRC Press is an imprint of Taylor & Francis Group, an Informa business

No claim to original U.S. Government works

Version Date: 20160226

ISBN 13: 978-1-138-07388-3 (pbk)
ISBN 13: 978-0-7546-7076-6 (hbk)

Visit the Taylor & Francis Web site at
http://www.taylorandfrancis.com

and the CRC Press Web site at
http://www.crcpress.com

Contents

List of Figures

List of Tables

Acknowledgements

The theme of this book was suggested by Professor Lennart Melin (Department of Psychology, Uppsala University), who also contributed ideas and feedback to several of the chapters.

Dr Lisa Dorn (Cranfield University) has also been an important discussion partner regarding many of the ideas presented here, some of which we have published joint papers upon.

Also, my collegues at the Department of Psychology at Uppsala University, who for a decade have been an inspiring scientific background for my various endevours. Amongst all these people, the various librarians who have supplied me with a steady stream of obscure references deserve special mentioning; Siv Vedung, Hans Åhlén and Karl-Oskar Göransson.

The text was language-edited by Charlotte Frycklund, Erna Magnusson and Lisa Dorn.

The following persons have kindly supplied requested information and/or papers upon personal contact, which is gratefully acknowledged:

> Professor Winfred Arthur Jr, Texas A&M University.
> Professor Russell A. Barkley, Medical University of South Carolina.
> Janine McLeod, RTA Library, Roads and Traffic Authority.
> Professor Cynthia Owsley, University of Alabama.
> Gillian Reeves, Parliamentary Advisory Council for Transport Safety (PACTS).
> Dr Jean Wilson, ICBC.
> Research Analyst Ming Fang, ICBC.

Also, the following persons have kindly provided feedback on various aspects of the manuscript, which is gratefully acknowledged:

> Professor Karel Brookhuis, Delft University of Technology.
> Professor Theresa Kline, University of Calgary.
> Professor Cynthia Owsley, University of Alabama.
> Dr Loren Staplin, TransAnalytics.
> Professor Heikki Summala, University of Helsinki.
> Professor Ola Svenson, Stockholm University.

The acknowledgement of the contributions of these researchers does not imply that they agree or disagree with the views stated in the book

The language-editing of this work was supported by the Swedish Research Council.

Introduction

It is a very common shortcoming of accident research that the effort expanded and the conclusions reached are not justified by the quality of the data employed. Haddon, Suchman and Klein, 1964, p. 85.

This is a book about assumptions. Within any human endeavour, there seems to exist a shared base of thoughts about how things work, thoughts that are seldom explicitly discussed, but are instead accepted as proven beyond any need for corroboration. This seems to be the state in science too, where knowledge accumulates and facts are added to an ever-increasing base of knowledge, but where some basics are accepted by most without discussion.

For research into individual differences in traffic safety, such a basis of accepted 'facts' seems to have come into existence during the last decades of the twentieth century, regarding certain methods and research findings, despite that this topic and its related research areas are less than a hundred years old, and thus still in their infancy. Globally, we are still far from safe in traffic, and the contributions to safety from research on individual drivers (as opposed to that on roads) to safety are only moderate to date. For example, there is no evidence that any practically useful method for selection of safe drivers has been constructed.

In this book, it will be argued that the weak results from this discipline are to a large degree due to the assumptions made within the field, because they are not only often unfounded, but have also led us astray in our search for knowledge about how drivers behave and cause accidents. This book will show that many studies undertaken are simply misdirected, and the results erroneous, due to (mostly unpronounced) beliefs about some of the basic facts of traffic safety, and the research methods used.

In the discipline of traffic safety research, there is little or no discussion concerning the methodology of individual differences. Sometimes known problems, which are mostly not very serious, are dutifully mentioned, for example, restrictions on randomness of samples etc. The implications of even these problems are then totally disregarded when it comes to drawing conclusions. Indeed, the weightier problems have only been mentioned by a few authors (and purely methodological papers in traffic psychology are extremely rare[1]), and their advice seems to have gone largely unheeded. Most researchers simply have a blind faith in the methods used, without any real knowledge of the relevant research concerning their validity. Assumptions rule, without being acknowledged, and often probably without being known to be assumptions. This book was written to make visible some of the common assumptions in traffic safety research (some of these are probably held

---

1   The exceptions include af Wåhlberg (2003a) and Sundström (2008), see also Bates and Blakely (1999).

by practitioners too), and to gather the evidence regarding their features, to enable traffic safety researchers in the future to avoid some serious pitfalls.

The present work can also be seen as an analysis of what traffic safety is considered to be, and what it ought to be, when undertaking research, or implementing 'safety' measures. As will be shown, there is little agreement about what constitutes safety or risk within the traffic safety research community, something that becomes very apparent when you look for and analyse certain parameters. There is no reason to think that practitioners within traffic safety are any less heterogeneous in their thinking and beliefs than the researchers, which might be a part of the explanation of why we have not managed to eradicate road traffic crashes.

The present text concerns itself with some research methods and results in traffic/transport safety from the individual point of view, so-called individual differences. This means that the individual is focused upon, whether as a single member of a population of road users, or as the culprit/victim in an accident. The individual road user is always the unit of analysis in the reviewed research. Furthermore, the main interest is upon the rarely mentioned problems with the methods used in this area of research. These might have a strong impact upon the conclusions drawn from a study undertaken with a certain method. Whenever relevant research is known, it is reviewed and analysed, and when none is known, there are sketches made about what type of studies need to be undertaken.

The main aim of traffic safety research is the prevention of road traffic crashes. As motorized vehicles are involved in the bulk of crashes, very little attention will here be given here to bicyclists and pedestrians. The possibility of generalizing to these groups of road users from the material presented is unknown, but the interested reader can probably easily see what similarities there are between methods and results for these different road user groups.

Subjects Covered

The main topic of this book could be said to be road traffic accidents. These are discussed from the viewpoint of how to categorize them, how to measure them, and how to use them as an outcome measure etc, all with the overall aim of understanding the relations between driver behaviour and characteristics, and the drivers' collisions. In the present perspective, crashes are intimately intertwined with driver behaviour, and the subject could therefore equally be said to be what drivers do in traffic, and how research is undertaken, and should be undertaken, to uncover these associations. In short, these are the topics of the chapters:

Chapter 1: Traffic Accident Involvement Taxonomies. In many ways, this is an introduction to several of the other chapters, because many of the problems of categorization of crashes and how this tool is to be used in research will surface again in the later analyses.

Chapter 2: The Validity of Self-reported Traffic Behaviour Data. Self-reports are the most common tool used by traffic researchers, and the importance of validating the measures used is therefore paramount. Here, this topic is discussed in terms of validation of variables, the possibility of spurious results and the argumentation of self-report researchers.

Chapter 3: Accident Proneness. Whether drivers are differentially dangerous over time, that is whether there is a stable disposition to cause crashes has been a debating point for a hundred years. The data gathered and analysed in this chapter proposes a much more central part for individual disposition, while also recognizing the environmental factors.

Chapter 4: The Determination of Fault in Collisions. As culpability for collision involvement is argued to be the most important way of categorizing crashes, the empirical evidence regarding how fault is determined and what effects this has on research results are presented.

Chapter 5: The Accident-Exposure Association. Controlling for exposure is one of the basics when it comes to predicting crash involvement. However, the results from studies on the association between these variables are rather peculiar (as are some of the interpretations of these results), and alternative explanations to the results are presented.

Chapter 6: Constructing a Driving Safety Criterion. Categorizing crashes is one thing, but what do different researchers do with these data when they want to use crashes as an outcome variable? It is shown that the number of methods for arriving at the criterion variable is bewilderingly large, and that no knowledge exists as to the effects of using different methods.

Chapter 7: Alternatives to Accidents as Dependent Variable. The use of so-called proxy variables (parameters that are thought to be suitable replacements for crashes as criteria) is presented. It is argued that most proxies are very poorly validated, and that their usage in individual differences research really only rests upon accepted tradition, 'common sense' and logical errors.

Chapter 8: Case Studies. A few choice papers are scrutinized in detail, with the aim of highlighting the methodological problems identified in the previous chapters.

Types of Analyses Used

In the present book, a quantitative approach (meta-analysis) was used in reviewing the available data whenever possible and useful, because the evidence was gathered from published research. However, the type of quantification differs from what is common in most meta-analyses, where the size of an effect in the population is usually the value sought. Instead, the goal of the present text was most often to investigate how the effect sizes varied with other values, like the mean in the samples, or effect sizes of different predictor variables. This kind of calculation is very uncommon (the only previous instance of its use that has been found is by

Struckman-Johnson, Lund, Williams and Osborne, 1989), and some descriptions of the general logic are therefore necessary.

It is often remarked by traffic safety researchers that the effects in their study were probably restricted by low variance. The meaning of this statement is that there is a true association in reality, but that the value we find in a study will necessarily be lower than the true one, if one of the variables (in this case collisions) does not vary much. This can be exemplified with a study of drinking; if we only measure how much alcohol the subjects drink during a day, this will vary between zero and maybe ten drinks. But if we measure for a year, the range of drinks will be from zero to thousands. From the latter, we can with a greater certainty predict who will become alcoholic (which is the same thing as the strength of the association between amount of drinking and later dependency). This effect is due to a lot more information being available in the latter case, and increased variance. If all subjects had one drink a day, we could not predict anything from this information, because there would be no difference between them (that is we would not have a 'variable').

The importance of this mechanism in traffic safety research is this: The effect sizes found when individual differences in collision record are predicted will to some part be decided by how the accident variable has been constructed (in terms of time period and the overall risk in the population), which determines the mean and standard deviation of the sample. Thus, studies that investigate the same subject, say the effect of personality on accident record, will have slightly different results, due to the different variances of their criteria. If this is seen as a variable, it can be used for a new type of meta-analysis.

This principle is, for example, the basis for the calculations in Chapter 3, where it is shown how the sizes of the effects (correlations) are strongly associated with the means of the samples (the mean is usually highly correlated with the variance, and more often reported, and therefore used in these calculations). A scatter plot of these two variables shows a fairly linear trend, and it can therefore aptly be described by a correlation or regression coefficient and the formula for the regression line. In effect the correlation in each sample/study is correlated with its mean.

Such a method may seem unorthodox but it is simply a matter of describing the trend in a number of data points. Whether these represent, for example, the height and weight of humans, or the correlations and standard deviations in some samples, does not really matter. The numbers may actually be exactly the same, and so will the output; a quantitative description of the association between the variables chosen.

Another use of the principle of calculating the association of an effect size with other data is when two different dependent variables are used in parallel to be predicted by the same set of independent parameters. If the two outcomes are considered to be measures of the same thing, for example, traffic safety, they should also yield the same pattern of associations with the independent variables, that is if traffic violations are highly correlated with exposure, then so should accidents be, if both these are thought to be outcomes of (the same) unsafe behaviour in traffic. If there are a fair number of other variables correlated with these two in a study, it is possible to take the correlations (or whatever statistic is reported) and correlate these values. Again, the correlation coefficient simply describes the size of the association between the sizes of the numbers, although this time these are both types of effect sizes. However, it should be noted that in this type of calculation, the regression line equation does not contain any information of interest, because what is investigated is not a trend in any particular direction, but only the similarity of two variables.

With this method, trends in published research can be analysed. The analysis of variance variation and effect similarity will here be called the secondary correlation method. The reason for doing this kind of calculation is that it can bring out effects that are not visible within a single study (variance variation), and that it describes the similarity between different criterion variables, where it is most often not noted that there are large discrepancies.

Also, some regular calculations of mean effects over studies will be undertaken, mainly aiming to bring out the differences caused by using different methods in the studies. As with the other methods, the reason for this is a belief that there are systematic effects of methodology in the published traffic safety research, effects that are not recognized today.

Comprehensiveness of the Data

The present book is based upon a fair amount of meta-research; extensive database searches and years of reading the safety literature have been undertaken to unearth relevant papers. However, many of the concepts where data have been sourced are often not the main themes of research undertaken, but are seen as unproblematic side issues by authors, and are therefore very hard to find. The reviews and meta-analyses of this book are therefore in several places probably far from all-encompassing. Therefore, the present analyses should not be taken as based on all available research, but on what has actually been found during a fairly comprehensive search of the literature, mainly in terms of journal papers. How much error is introduced by this shortcoming cannot be known. However, any reader who wants to point out a relevant paper that is missing is welcome to do so.

Terminology

The present book uses the standard vocabulary of traffic research (including statistics), with a psychological slant, and should thus be easy to understand for those knowledgeable about the scientific literature in this area. However, the very central term 'accident' might need some comments. At times, one does come across the view that 'accident' is not a good term to use (for example, Langley,

1988), because it has connotations of randomness and inevitability. However, as should be apparent from the present text, the meaning intended is that of unintentional adverse traffic events, usually with some sort of resulting damage. This is sometimes referred to as a crash. However, 'crash' may at times seem to be a bit too strong an expression. Does this also encompass small dents on your car? 'Collision' on the other hand would seem to be fairly neutral, just describing that two objects have made contact. But does this include someone going off the road, without hitting something? How about a bicyclist falling off his vehicle? In the present work, several terms (accident, crash, collision, mishap and incident) will be used interchangeably, all denoting unforeseen traffic events which cause damage to objects, people or other animals, by the movement of road users. All these forms are usually only short for 'involvement in X', that is it is not the event as such that is intended, but the actions and results of one of the involved parties.

In several instances, data has been taken from tables and figures in various papers and re-calculated, that is the value presented here was not included in the paper referred to. In these cases, it is stated that 'data was taken from table 1' or similar. This means the table in the original paper, not any of the tables in the present text.

Aims and Audience of the Book

The present work is intended to create a discussion. Do we need to be more strict about our research regarding individual differences among road users? Do we need to re-appraise things that we now take for granted as true within traffic safety? Can we actually trust the bulk of the traffic safety research literature? These kind of questions lead to some speculations about the true nature of certain variables, because sometimes the available data would seem to contradict common views among traffic safety researchers. In general, this book aims to criticize a major part of the research into individual differences in traffic safety of today as unreliable, due to its methodological shortcomings, and challenging the criticized researchers to come up with better evidence of their claims.

This book is intended for a fairly specialized audience, with at least a basic knowledge of (mainly psychological) research methods within traffic safety, as well as statistics. However, it might also be used as a textbook for specialized students, and transport professionals, as well as editors and reviewers of journals when making decisions about accepting or rejecting papers. For these audiences, as well as for researchers, the aim is to challenge what often seems to be unconscious beliefs (assumptions) about traffic safety, and hopefully to make them visible to those who adhere to them. In the end, we need to re-think a lot of traffic safety, because the present picture is based to a large degree on faulty research.

Anders af Wåhlberg

This book is dedicated to Professor Lennart Melin (Department of Psychology, Uppsala University, Sweden), who suggested the theme

Supervisor, mentor, friend
and, above all,
scientist

Chapter 1
Traffic Accident Involvement Taxonomies

Introduction

> Fortunately, accident researchers do possess a reasonably good criterion - actual
> accidents or collisions ... McGuire and Kersh, 1969, p. 20.

Most researchers would probably agree that traffic accidents are not a homogenous group of incidents (as argued by Babarik, 1968; Barrett and Thornton, 1968; McBain, 1970; Golding, 1983; Ball and Owsley, 1991), and that they could be usefully classified into sub-groups, that is a taxonomy could be constructed for them, where the categories would have different causes or other defining differential features. These categories of incidents could thereafter be used as outcome variables in studies on individual differences in traffic safety. In general, it would seem to be as stated by Whitlock, Clouse and Spencer (1963); 'It frequently occurs that an injury is not the result of an accident behaviour, but the injuries the psychologist can hope to predict are only those which do result from accident behaviors.' (p. 35).

Somewhere here, however, the agreement stops. Researchers, insurance companies, the police, hospitals; everyone and anyone has their own system. The reasons for this may partly be practical; different groupings may be useful for different purposes, for example, junction accidents in conjunction with older drivers (for example, Owsley, Ball, Sloane, Roenker and Bruni, 1991) or shunts with sharp braking (for example, Babarik, 1968). However, it might also be suspected that taxonomies as a methodological tool has not been sufficiently developed within traffic research, and maybe especially so for traffic psychology. The number of studies using no categorization whatsoever is large, while those actually using some variation of this method are few, as will be evident in this chapter.

The present chapter concerns traffic accident taxonomies constructed from the viewpoint of the individual, that is in relation to the (presumably) accident-causing behaviour of drivers, which is a psychological typology. Four other main groups of systems can be discerned; those that focus on the vehicle (for example, Stein and Jones, 1988), the accident as such (for example, Preusser, Williams and Ulmer, 1995; Sohn and Shin, 2001), injuries (for example, Farmer, Braver and Mitter, 1997) or the road (for example, Retting, Williams, Preusser and Weinstein, 1995; Retting, Weinstein, Williams and Preusser, 2001; Ivan, Pasupathy and Ossenbruggen, 1999). However, there is often some overlap, for example, when driver characteristics are used in studies about vehicles (although these 'characteristics' tend to be rather crude, like age and sex). It can also be noted that the non-psychological taxonomies often use the accident as the unit of analysis, while a psychological approach to accident taxonomies is defined by the use of the individual as unit of analysis and source of data (in the sense of a characteristic). A relevant question in this kind of framework might be; is there a difference between drivers, using this psychological variable, that is predictive of which type of accident the driver is involved in?

It can be stated from the very beginning that very few researchers have been interested in a comprehensive approach to psychologically based accident involvement taxonomies. Among the very few examples can be mentioned McGlade and Laws (1962), Panek and Rearden (1987), West (1997) and to some extent af Wåhlberg (2002b; 2004a). West's system is the only one that seems to have had any significant impact within traffic psychology (it has been used by, for example, Parker, 1999; Parker, McDonald, Rabbitt and Sutcliffe, 2000; Fergusson and Horwood, 2001). Otherwise, researchers make up their own systems (or divisions) as they go, often as a mixture between the different types of taxonomies specified above.

So, why do we need traffic accident taxonomies? Within traffic psychology, where much research is about predicting accident involvement (for example, see reviews by Lester, 1991; af Wåhlberg, 2003a), the importance lies in what kind of behaviour leads to various kinds of accidents, and the use of this behaviour for prediction (or some other variable which causes differences in behaviour, like sex). Thus, single vehicle accidents could be suspected mostly to be due to excessive speed, while crashes in junctions more often would seem to be caused by improper attention. If a researcher wants to ascertain whether a tendency to drive fast is indeed correlated with vehicle accidents, should they use all the subjects' collisions as dependent variable? Not unless it is a study with multiple predictors, because it is improbable that the same drivers would tend to have both single and junction crashes (and these thus being inter-correlated), as the first is basically a problem of the young driver (Brorsson, Rydgren and Ifver, 1993), and the latter of the older. If non-speed crashes are included in the criterion variable, they will therefore probably just add error variance, making it more difficult to reach statistical significance.

In other words; accidents can be caused by many behaviours, and a great many variables are therefore associated with them (Peck, 1993). This is but one reason for the use of divisions within the accident phenomenon, many more can be presented. However, basically they are all about the non-homogeneity of traffic incidents. As in life in general, there are many ways to be injured or die in a traffic accident.

In this chapter, the use and (to some extent the) non-use of traffic accident taxonomies by traffic safety researchers will be discussed. What possible dimensions of traffic accidents are there and what is the evidence for them as useful groups? How and why are categories constructed and used? What is their theoretical basis, if any, and do they have any impact on research where they might be of use?

Possible Dimensions of Traffic Accident Involvement Taxonomies

Culpability[1] and Similar Constructs

> Failure to control for the factor of culpability in accident research is a serious methodological error which can obscure meaningful associations or otherwise lead to erroneous conclusions. Banks, Shaffer, Masemore, Fischer, Schmidt and Zlotowitz, 1977, p. 13.

Possibly the most important split of accident involvements for individual differences research pertains to the responsibility for the incident. The main argument for the use of culpability as a categorizer is that some accidents can be said to be independent of the behaviour of at least one of the drivers, for example, being shunted when having stopped for a red light (in a slow and controlled manner, at least). The important thing is that no psychological variable of the shunted driver will be able to predict that kind of accident[2] (if the stopping of the first vehicle is abrupt, the situation is different, see Babarik, 1968).

How then, should culpability be defined? In theory, there should be a clear line of demarcation between incidents, which have been caused by a driver, and things that have just 'happened' to him, that is been caused by some other road user. However, such a statement will just beg the question; what then is the definition of 'caused'? And to muddle things even further, it is easy to envisage that there are degrees of responsibility, with the clear cases of single accidents (not due to unforeseen technical failure) fading over into shared responsibility. Where should the line be drawn?

Starting with the most extreme case, it can be argued that all accidents that happen to a driver are their responsibility, because they were there. If the driver had gone somewhere else, it would not have happened (at least not to this driver). But is this a satisfying definition? It would seem that there is indeed a difference between being pushed off a cliff by a stranger passing by, and falling while climbing the same cliff. What is needed is a definition that will entail this intuitively attractive split. Therefore, it is suggested that a culpable accident is where a driver did not really have any possibility of avoiding the incident, apart from being somewhere else (a baseline level of risk because of exposure is accepted), as stated in af Wåhlberg (2002b).

Considering the research on individual traffic accident prediction, and interpreting the various ways this is undertaken, there seem to be four main views among researchers; culpability (or similar constructs) is:

---

1   No legal meaning is intended with this term, only if driver behaviour caused a crash.

2   Unless amount of exposure can be said to be an individual differences variable, in the sense that some drivers prefer to be more mobile than others. This is possible but so far untested.

1. A basic factor, which needs to be included in the method, and possibly, investigated as a factor, but not really discussed.[3]
2. A difficult notion that needs to be investigated and/or included in accident prediction methodology (for example, af Wåhlberg, 2000; De Raedt and Ponjaert-Kristoffersen, 2001).
3. Not of any consequence for research on accidents, and thus not included in the analysis or discussed.[4]
4. Found inconsequential in empirical studies and therefore left out (Shaw, 1965).

Unfortunately, there is no consensus among the researchers in groups 1 and 2 as how to define culpability, in an optimal or even standardized way. Actually, there is very seldom any definition of culpability included in studies using this concept, or any number of other terms that presumably mean something similar, like 'preventable' (Parker, 1953), 'primary vehicle' (Munden, 1967), 'chargeable' (Gumpper and Smith, 1968), 'to blame' (Quimby, Maycock, Carter, Dixon and Wall, 1986), 'responsibility' (Hartley and El Hassani, 1994) and 'fault' (Dobson, Brown, Ball, Powers and McFadden, 1999). Fairly often, it is not even the researchers who make the definition and coding, but police officers, transport companies (for example, Parker, 1953) or crash investigation teams (see Chapter 4 for more information).

Furthermore, almost nobody discusses why they have chosen to use culpable crashes as outcome variable. Most have just by-passed the problem, probably on theoretical grounds although these are most often not stated, for example, Panek and Rearden (1987), De Raedt and Ponjaert-Kristoffersen (2001), af Wåhlberg (2004b), and Wilson, Meckle, Wiggins and Cooper (2006).

So far, the discussion has been largely theoretical. What evidence is there that culpability for accidents is actually a factor when it comes to optimal prediction of accident involvement? And how can this be determined?

The basic method is fairly evident and similar for all types of categorization; the type of accident of interest should be more strongly correlated with the predictor than the rest of incidents (af Wåhlberg, 2008a). However, it should be remembered that this might lead to an uneven split, with considerably more variation in one sample, leading to a stronger effect for this group. If the category used is the larger part of accidents, then the comparison is probably better made between this

---

3   For example, Cation, Mount and Brenner, 1951; Gumpper and Smith, 1968; Quimby and Watts, 1981; Garretson and Peck, 1982; Avolio, Kroeck and Panek, 1985; Ball, Owsley, Sloane, Roenker and Bruni, 1993; Hartley and El-Hassani, 1994; Goode et al., 1998; Vernon, Diller, Cook, Reading, Suruda and Dean, 2002; Legree, Heffner, Psotka, Martin and Medsker, 2003.

4   For example, Parker, Manstead, Stradling and Reason, 1992; Decina and Staplin, 1993; Gregersen, 1994; Parker, Manstead and Stradling, 1995; Cliaoutakis, Demakakos, Tzamalouka, Bakou, Koumaki and Darviri, 2002.

and 'All accidents'. It can be said that if 'All accidents' yields a stronger effect than the sub-group (preferably after exposure has been controlled for), then the categorization does not work.

The available evidence regarding culpability as a categorizer is mainly of the type described above; comparing 'at-fault' incidents with 'All' as dependent variables. Thus, Cation, Mount and Brenner (1951) found only 'avoidable' accidents to be significantly associated with their predictor (reaction time), while Garretson and Peck (1982) found that fatal accidents judged by the reporting police officer to be the drivers' fault differed (by higher values) from other fatalities on a number of variables, mainly concerning registered legal violations and licence suspensions. The culpable drivers were also very much more likely to have been drinking, but the authors (very rightly) cautioned that there may be a strong influence on this factor on the judgement of the officer of the responsibility for the crash, regardless of the actions of the driver (as shown experimentally to be the case for eye-witnesses by Köhnken and Brockmann, 1987).

Similarly, when comparing 'at-fault' accidents with all available, several researchers found the first to be better predicted by variables such as traffic offences (Rajalin, 1994), driver performance (Gully, Whitney and Vanosdall, 1995), personality (Arthur and Graziano, 1996), various questionnaires (Lajunen, Corry, Summala and Hartley, 1997), various psychological tests (Parker, 1953), parent/child accident record (Wilson, Meckle, Wiggins and Cooper, 2006) and possibly self-reported lapses, errors and violations (Dobson, Brown, Ball, Powers and McFadden, 1999, these results were only mentioned in passing as a tendency). See also Tables 4.5 and 4.6 in Chapter 4.

Using the active/passive distinction of West (1997)(see the next section), Fergusson and Horwood (2001) found the former to be more strongly associated with cannabis use. Similarly, McKnight and Edwards (1982) found that in one of their two treatments, collision involvements with convictions (which should be a sub-group of culpable ones) differed more strongly between groups, while Banks, Schaffer, Masemore, Fischer, Schmidt and Zlotowitz (1977) reported that legally culpable drivers had three times more violations on record than drivers who had not been found to be responsible for their accidents. It should be noted that the comparison of culpable/non-culpable is different from the culpable/all accidents comparison, as the non-culpable category is usually at a disadvantage, due to lower variance in the latter.

On the other hand, some researchers have not found the expected difference, or they have found the opposite, on this parameter; Quimby, Maycock, Carter, Dixon and Wall (1986) for perceptual variables; Arthur, Strong and Williamson (1994) for visual attention (two out of three), Goldstein and Mosel (1958) for violations, Leveille, Buchner, Koepsell, McCloskey, Wolf and Wagner (1994) for medications, and McBain (1970) for a number of variables.

The evidence concerning culpability is thus not totally conclusive, although the differences in the overall results are probably, at least in part, due to slightly different definitions and populations used, as well as other methodological peculiarities (see Chapter 4, where this problem is further analysed). However, at times it is a difference of some practical importance. Despite this, the use of culpability as categorizer would seem to have lessened with time (if the active/ passive (West, 1997) distinction is seen as not being part of the culpability group); in the review af Wåhlberg (2003a) the mean publication year for studies using this split was eight years lower than for the other studies. Also, only about 17 per cent of studies used the culpability split (25 per cent if active/passive was included).

It can also be observed that, most often, culpability is used as a dichotomous variable (for example, Dobson, Brown, Ball, Powers and McFadden, 1999); few have used a more differentiated scale (the exceptions include Kahneman, Ben-Ishai and Lotan, 1973; Hartley and El Hassani, 1994; and possibly Nagatsuka, 1970). The use of a rating scale with several steps in it could possibly increase inter-rater agreement (and thus validity) and statistical power. On the other hand, depending on the method used, it could also result in a lessened power due to the splitting of samples, that is, if the interest was only in a particular part of the scale. The various ways of using accidents as dependent variable (not necessarily with regard given to culpability) will be further discussed in Chapter 6.

It may also be noted that a reason for using culpable accidents only is that they should have a higher reliability, that is, random effects on collision records of drivers should be less for this category. However, the research into whether this is really the case will be covered in Chapter 3 (the stability of accident record over time), and it may be sufficient to say here that this question is not yet resolved, as the available data contain a strange anomaly.

One possible reason for the not very clear differences on various psychological predictor variables when using culpability as a categorizer is simply that the criterion may be wrong. As the accident variable is very sensitive to error, due to the low variation, and that culpable and non-culpable accidents are probably slightly correlated within drivers, due to amount of exposure, a harsh or a lenient coding may eradicate the existing differences from a sample. This problem will be investigated in Chapter 4. However, Robertson and Drummer's (1994) scoring method is recommended for culpability judgements, as well as the classification into degrees of culpability used by Terhune (1983), although it is recognized that these methods need validation. How that can be undertaken is also presented in Chapter 4.

Active-Passive Accidents

> The active-passive distinction provides a reasonably objective measure of responsibility. Roberts, Chapman and Underwood, 2004, p. 213.

One of the very few traffic accident taxonomies in existence with an individual differences slant seems to be the Accident Script Analysis approach by West

(1997). It could be discussed under the heading of culpability, or type of behaviour preceding accident, but due to its popularity and a number of peculiarities, it has been given a section of its own.

Basically, accident script analysis evolved in an empirical way from descriptions of collisions given by the drivers themselves. It mainly concerns the relative movements of the vehicles, that is, the behaviour of the drivers. The important difference between this system and others is the active/passive notion (the other categories are physical descriptions of movements before the crash etc), and here also lie the main problems. Basically, in each two-vehicle crash, one driver is designated by the researcher as the 'active' part, and the other as 'passive', based upon the self-report of a driver. The assignment of these labels would seem to be mainly dependent upon the speed of the vehicles, with the faster one seen as active. Curiously, if a driver pulls out in front of another in a junction and is hit, he is usually designed as passive, even if the other had the right of way. Here, the difference versus a culpability construct would seem to be clear.

West's taxonomy has been criticised in detail before (af Wåhlberg, 2002b), and only the main points will be reiterated here:

1. The active/passive distinction has been given various definitions which are not quite compatible (compare Elander, West and French, 1993; Parker, West, Stradling and Manstead, 1995; West, 1997; West and Hall, 1997; West and Hall, 1998). It seems that in these publications there is a vacillation between a legal culpability definition and a behavioural one, while Horwood and Fergusson (2000) and Fergusson, Horwood and Boden (2008), used a combination of both, which in the last study would seem to be very similar to culpability.

2. The reasons for constructing the active/passive categories are only ever hinted at, and mainly seem to be about avoiding socially desirable responding in questionnaires (West and Hall, 1997), and constructing a system that is easy to use when gathering data (West, 1998). However, no one seems to have tested whether people do indeed report more truthfully with this system, or if the system is easier to use than any other. The only stated reasons for the active/passive category are therefore hypothetical and lack empirical backing.

3. The assumptions about the feasibility of the system are at times somewhat strange and/or lack empirical support. For example, it is said that '... the categories are sufficiently distinct that it would require complete fabrication of accident details by drivers to move an accident from its true category to another ...' (Parker, West, Stradling and Manstead, 1995, p. 573). The authors thereafter argued that if the respondents were unwilling to disclose the true details, they would leave the space blank or not return the questionnaire. Now, even if this was true (and no evidence was cited in support), the question of whether the drivers actually can remember the details of their crashes was not discussed. In fact, it is actually very likely that drivers will

report willingly but erroneously on their accidents, in numbers and details (af Wåhlberg, 2002a), see further Chapter 2.

Roberts, Chapman and Underwood (2004) claimed that 'The active-passive distinction provides a reasonably objective measure of responsibility.' (p. 213). No evidence was cited in favour of this position, and it was not explained why the active-passive distinction was used, if the aim was to distinguish culpability. Furthermore, it is hard to understand why the use of the active-passive categories should be more objective than the assignment of culpability straight away.

4. The system always denotes one driver as active and the other as passive. Although everyone is free to define the terms they use in their own way, this leads to an assigned difference that may not always be true. If one driver slams on the brakes on an open road and is hit by the one behind, is there a qualitative difference in their respective behaviours which can be assigned different terms which seem to imply different types of actions?

Curiously, before the Accident Script Analysis report (West, 1997) was published, the terminology was introduced in West, Elander and French (1992), but with a fairly different meaning. In this early version, active/passive was much more like the culpability concept. In fact, it can hardly be distinguished from that used in af Wåhlberg (2002b). It may therefore be logical that Clarke, Ward, Bartle and Truman (2006) actually equated these two concepts.

Anyway, in general, the accident script analysis of West therefore seems to be incoherent, inflexible and lack theoretical grounds as well as empirical support. Before any empirical backing of its main theses (whatever these are) can be shown, it should not be used for research. Unfortunately, it has been used a number of times (for example, McDonald, Parker, Sutcliffe and Rabbitt, 2000), mainly in conjunction with the Manchester Driver Behaviour Questionnaire (for example, Parker, McDonald, Rabbitt and Sutcliffe, 2000; Fergusson, Horwood and Boden, 2008).

Minor-Major/Severity

Given a crash, the use of alcohol or drugs shows no clear association with the severity of the accident. Smink, Ruiter, Lusthof, de Gier, Uges and Egberts, 2005, p. 431.

In classical accident proneness theory, no difference between minor and major incidents would seem to have been made (see Chapter 3), indicating that they were considered to be similar, and therefore correlated within individuals. However, this accepted view was challenged by Adelstein (1952), who, using data from the South African railway (1,451 work accidents of shunters), showed that there was no significant correlation, neither between severe injuries (> six days absence) and minor ones, nor between injuries and damage only accidents, for the same men. On the other hand, Farmer and Chambers (1926) did find correlations in the range of zero to .356 between major and minor accidents in various categories of workers.

Are minor traffic accidents caused and therefore predicted by the same variables (or drivers?) as major (severe) ones? This could also be phrased as; does a certain variable predict one category of crashes, but not another? Few researchers have used this kind of categorization,[5] but in line with the initial arguments about the possibility of different behavioural causes of accidents, this split is indeed worthy of some consideration, and a look at the evidence would seem to indicate that this question is far from resolved. One of the reasons for this is a lack of data correlating major and minor traffic accidents. The only results found so far are those of Dahlen and Ragan (2004), who reported a correlation of .47, and Dahlen and White (2006), yielding .42 (both studies used self-reported data). What type of conclusion such results support is uncertain. On one hand, the self-report method probably inflated the correlation, while the low variances of the variables probably deflated it. In the end, it is therefore not possible to know what these results mean.

However, the problem of correlation between minor and major incidents can also be tested by the use of the secondary correlation method, as there are a few studies that have used the categories of major and minor accidents as outcomes in parallel, as shown in Table 7.2. It can be seen in this table that in most cases, results were very dissimilar between variables, to such a degree as even to yield a negative secondary correlation in one case. This result seems to support the notion of different severity categories, especially as the data in this table was all self-reported and therefore probably artificially correlated.

However, the data available for secondary correlations are scarce, and so now we turn to other types of results. Concerning severity of crashes, Zhang, Lindsay, Clarke, Robbins and Mao (2000) found that a four-level categorization of injury-producing accidents yielded several differences between the resulting samples of drivers, while Smink, Ruiter, Lusthof, de Gier, Uges and Egberts (2005) found no really distinctive evidence for the proposition that drug-related crashes are more severe than others. If anything, these accidents were less severe. Also, Guibert, Duarte-Franco, Ciampi, Potvin, Loiselle and Philibert (1998) reported that their results were similar when using crashes with and without injuries.

Taylor, Chadwick and Johnson (1995) used the more fine-grained method of three categories; any accident, injury producing accidents and serious injury accidents. For the three medical variables used, the confidence intervals increased with severity, as expected. However, for two variables (anti-epileptic drugs and

---

5   The exceptions include: Davison, 1985; Matthews, Dorn and Glendon, 1991; Matthews, Desmond, Joyner, Carcary and Gilliland, 1996; Deffenbacher, Filetti, Richards, Lynch and Oetting, 2003; Deffenbacher, White and Lynch, 2004.

period seizure free), the odds ratios decreased, while for the last variable (warning of epileptic attack), they increased. This could be an example of differential prediction by severity, but as the effects were not significant, they should be treated with some caution. Maycock and Lockwood (1993) reported that damage-only and injury accidents varied differently in association with mileage, age and experience, but as this study contained mathematical modelling with several unspecified corrections (see also Chapter 5), this study is also hard to evaluate.

The use of fatal accident involvements as a categorizing factor is very similar to severity in terms of physical damage, and has yielded some results. Thus, Garretson and Peck (1981; 1982) and Helander (1984) found that drivers in fatal accidents had more deviant driving records than the general population. However, the latter also found that those with three previous accidents were even more deviant.

In summary, there appear to be insufficient results to refute or validate the existence of different severity categories of accidents, in terms of being caused by different variables or drivers. What can be concluded with certainty is that many researchers do restrict their accident variable to the severe category by using injuries as dependent variable (for example, Norton, Vander Hoorn, Roberts, Jackson and MacMahon, 1997; Jelialian, Alday, Spirito, Rasile and Nobile, 2000; Blows, Ameratunga, Ivers, Lo and Norton, 2005). If this is due to theoretical considerations, or simply to the fact that the researchers using this method are often from the medical profession, is not known.

Type of Behaviour Preceding Accident

> Because of the multiplicity of factors that influence accidents, it is very difficult
> to discern the effect of any particular driver characteristic. McKnight, Shinar
> and Hilburn, 1991, p. 226.

For many types of accident research, the human behaviour that preceded the incident is of prime interest and importance (as it is for the legal side), as exemplified in the speed collision example given previously. Despite this, accepted categories are lacking in this area, too, and the reasons for the development of each specific taxonomy are seldom stated.

For example, McKnight and McKnight (2003) wanted to compare the various accident antecedent behaviours of drivers of slightly different ages in the beginning of their driving careers to see whether the increase in experience led to changes. Despite being in general a well-conceived study and a well-written paper, the matter of the development of the behavioural categories was very shortly described, probably because the authors, as they stated, did not consider their classes as a taxonomy, but only a means of comparing behaviours over time/ experience. Although this is an acceptable reason, and the main aim of the study was met, it is an unfortunate state of affairs that almost every researcher uses their own classification, thus making data hard to compare between studies.

Should accidents caused by drivers under the influence of drugs/alcohol be considered a separate category? There are some peculiarities to these incidents that seem to set them apart (see the review by Zylman, 1974), like the very high probability of the drunk driver to get killed in a multi-vehicle crash. But is the fact that the driver was drunk not merely a distal factor, and the real importance lies in the actual behaviour preceding the crash? It could be argued that when studying the circumstances of a crash to determine responsibility, an error is an error, regardless of the possible intoxication of one of the drivers. If someone pulls out into traffic when he should have given way, does it matter whether he does that because he wants to show off to his friends, because he is very near-sighted, sleepy, hung over, or still over the legal BAC limit? A few researchers have used only drink driving crashes as outcome variable (for example, Elliott, 1987). This approach invariably runs into problems of identification, and comparative studies on predictive power seem to be lacking. It could be suspected that drivers who drive drunk cause traffic accidents even when they are sober, and the problem of outcome variable for alcohol studies would therefore very much seem to be a matter of what the aim of the study is, that is what one does want to predict.

At times, categorization according to sleepiness as the cause of accidents has been reported. Here, the effect for this sub-category was larger in the study by Cassel, Ploch, Becker, Dugnus, Peter and von Wichert (1996), in terms of per cent reduction of accidents, while de Assis Viegas and de Oliviera (2006) found that drivers with a high Body Mass Index (which is related to sleep problems, mainly snoring) had more sleep-related crashes than other drivers, but fewer accidents in total.

Another possible categorization is by whether accidents are seen as caused by excessive speed. Only one study using this method has been found; Horswill and McKenna (1999). The results were inconsistent. A video (simulator) speed test could predict speed-related accidents better than it could predict all crashes, while a speed questionnaire was more strongly associated with the latter.

Finally, Smith and Kirkham (1982) predicted that intersection crashes would be a more prevalent problem for drivers with low intelligence, and therefore used such incidents in parallel with 'All accidents' as outcome. The result did not favour the hypothesis, but the authors did not take differences in variance of the accident variables into account, so this effect cannot really be interpreted.

In summary, there is hardly any evidence at all regarding the feasibility of driver behaviour as a collision categorizer, although it is as times used. Given the troubles of ascertaining pre-collision behaviour, especially from later self-reports, it could be doubted that this method will actually be successful.

Single Versus Multiple Vehicle Crashes

Single accidents, that is, those not involving any other road user other than the driver of one vehicle, would seem to be a good candidate for a categorizer that could yield different results as compared to 'All accidents'. However, no research aimed specifically at this question has been found, and very little of anything that has a bearing on it.

Ivan, Pasupathy and Ossenbruggen (1999) found different significant predictors for single accidents as compared to multi-vehicle ones. However, this was not an individual differences approach, but a technical/engineering one, and the predictors were therefore about road layout, traffic density and similar factors. Still, this may indicate that there are differences between drivers too, as drivers probably react differentially to the road environment.

Similarly, Lenguerrand, Martin, Moskal, Gadegbeku and Laumon (2007) reported odds ratios for risk of being involved in an accident due to use of cannabis and alcohol (five levels), using three different methods of keeping exposure constant; case-control, quasi-induced[6] and single vehicle. The last method used the not responsible drivers from the quasi-induced sample as controls. It was found that for all six cases, the case-control and quasi-induced method yielded very similar results, while for the single-vehicle approach the values were all much higher. This would seem to indicate that the risk of having a single-vehicle accident is higher than having a multiple-vehicle one, when under the influence of drugs. Such a result does make sense, given the prevalence of driving under the influence at night, when there is less traffic, but could also be interpreted as a result of a preference of intoxicated drivers to avoid busy streets, and/or taking greater care in such areas. However, regardless of the reason for the difference in risk, it can be tentatively concluded that single-vehicle accidents seem to be, at least for some predictors, different from other collisions.

Theoretical Categorization

> Day-time accidents should be excluded when attempting to associate night vision scores with safe driving ability. Davison, 1978, p. 131.

Very few, if any, traffic accident taxonomies are the result of explicit theoretical considerations (West's accident script analysis is not grounded in theory, even though the active/passive distinction could possibly be developed into something useful). One possible candidate is the hierarchical theory of driver behaviour of the traffic psychology group of University of Turku (for example, Laapotti,

---

6   Induced exposure methods use non-culpable accidents or similar categories as proxy variables for exposure. This is discussed further in Chapter 4.

Keskinen, Hatakka and Katila, 2001). In this model, four levels of behaviour[7] are assumed, at increasing levels of abstraction; operational, tactical, strategic and life goals. Next, the authors seemingly intended to categorize accidents according to at which level a failure had occurred and caused the incident, with the aim of testing whether sex differences in accident categories had changed over the last decades of the 20th century (Laapotti and Keskinen, 2004). However, it was not apparent from the paper exactly how the categories used were related to the model. Actually, the authors did not seem to be certain about this themselves: 'The accidents were classified into four types ... These types are seen to represent different levels of driving behaviour according to the hierarchical model ...' (p. 579). These sentences would seem to state that each level had its own category of accidents. However, thereafter the authors say 'The two lower levels of driving behaviour (vehicle handling, control of traffic situations) are present in: ...' after which they define the categories. Furthermore, on the next page, it was explicitly stated that the categories are not exclusive, and that all levels of failure may be involved in causing an accident.

Now, the final statement may be perfectly reasonable, as most researchers would probably agree that there are multiple reasons for behaviour and accidents. But it also wipes out the connection between the hierarchical model and accident types. So why is the model invoked into the study at all? It did not really do any good. Instead, it would seem that the results of the study (mainly no differences in differences over time between sexes) were vaguely invoked as evidence in favour of the model. But no additional data was gathered measuring the levels of the hierarchy, and the collisions were not categorized in any way according to the model. Therefore, the model was not really necessary for the study at all, but instead a hindrance, mainly because it demanded some space, but also because of the confusion about the relation between accident categories and the hierarchy.

Concerning the Laapotti and Keskinen (2004) paper there is also a possible problem that could have an influence on the conclusions about sex differences in the accident data. There may have been a coding bias, for example, concerning the culpability variable. Included in this study were only incidents that the driver was held at least partly responsible for by the state investigation team. But how impartial were these teams when it came to establishing guilt by males versus females? Considering the evidence for how easy it is to influence people's judgements of culpability with other variables (for example, Köhnken and Brockmann, 1987), it is equally easy to suspect that something like this might be in operation for sex too.

There is thus no evidence, or indeed any stated reason, for a classification of crashes by the hierarchical model, which is a further indication of its lack of theoretical properties. No testable predictions seem to have been extracted from

---

7   Strictly talking, these levels are not levels of behavior (as claimed by Laapotti, Keskinen, Hatakka and Katila, 2001); only the first is actually about observable actions, the rest are constructs that are assumed to influence the behavior of the driver.

this system, regarding crashes or anything else, and it can therefore not be used for such ends such as those discussed in the present chapter, at least not in its present state.

Discussion and Conclusions

> The unreliability of the criterion may in some measure be due to the fact that accident data are not uniform, and that by merely including for study all accidents ... irrespective of causes or the manner in which they occurred, one is in effect collecting a 'hotchpotch' of events which are by no means homogenous, or representative ... Arbous and Kerrich, 1951, p. 382.

Overall, it would seem that the use of categorization of accidents as a methodological tool in traffic psychology is heterogeneous, under-developed and largely ignored. A paper like that of Cercarelli, Arnold, Rosman, Sleet and Thornett (1992), which discussed the proper type of motorcycle accident to use as outcome variable in different studies, is very rare indeed. In a few other instances, the problem is mentioned, but only in passing (for example, McBain, 1970).

This is somewhat strange, as there seems to exist a considerable activity in neighbouring research areas (the main varieties of taxonomies identified in the introduction, and cross-overs), at least when it comes to using categorization as a research tool, because there does seem to exist classification systems for accidents in general, which are used in many studies (see the review by Lortie and Rizzo, 1999), while, for traffic accidents, this is very much lacking.

When a question on taxonomy is asked, for example, within biology, a fervent discussion is likely to break out. Indeed, various taxonomies form an important part of that discipline, being indispensable working tools. Within traffic research, this is definitely not so. On the contrary, most psychological studies on individual differences in traffic safety do not use any kind of categorization of accidents at all. Unfortunately, no explanation for this state of the art can be found in the traffic safety literature; just as taxonomies themselves, the discussion of their use and non-use is largely non-existent, and it may be observed researchers who would really need to discuss this (those who do not use them) do not.

Today, the largest set of available evidence concerns culpability as a categorizer, and most of it would seem to indicate that it is indeed a useful parameter. So, why do not all traffic researchers use the culpability parameter in their studies, given the results reviewed above? Some methodological problems can be identified; maybe these can explain why the results have not always been clear, and why some researchers do not think it is a worthwhile undertaking to ascertain responsibility for accidents.

A basic problem in the use of the culpability variable is the widespread use of self-reported accident data. Given the strong tendency of subjects to misrepresent their road mishaps (see Chapter 2), it is probably not very useful to try to ascertain this parameter in data such as this, although some have tried (for example, Dobson, Brown, Ball, Powers and McFadden, 1999; De Raedt and Ponjaert-Kristoffersen, 2001).

It can also be claimed that it is not really possible to ascertain culpability, given the methods (or lack thereof) that we have today. This problem will be covered in Chapter 4, but it can be noted that if one does not try, there seems to be little use in claiming that it does not work (for example, Verschuur and Hurts, 2008).

Are there different 'accident pronenesses' (as suggested by Adelstein, 1952), that is, are some people destined to have small mishaps, and others big ones, but very few to have both? This is not necessarily so, because another explanation is compatible with the evidence, although it would need to be specifically investigated for some support. It could be that an initial proneness may exist (say, when a driver acquires a licence). However, this proneness (risky behaviour) will be modified by the accidents that happen to the driver (as discussed by Adelstein, 1952), with small incidents having very little impact (and thus being measurably stable over time), while major crashes will have a large influence, and thus in practice prohibit the stability of a severe crash variable. However, this mechanism would require that drivers could actually distinguish behaviours that might lead to minor and major crashes, something that has not been investigated. Indeed, it can be questioned that people even know what behaviours lead to any type of mishap.

It is therefore suggested that the reaction (ensuing behaviour change) of a human to an accident is a function of the severity of the mishap and the self-assigned degree of culpability, where severity in its turn is a function of amount of damage and injury. However, the exact shape of the severity function probably differs between people and cultures, although most would probably consider injuries to be more important than other damage. Furthermore, the effect will diminish over time with approximately the same strength for different crashes, meaning that a severe incident will have a longer-lasting effect.

Some of the difficulties in using categorizing in accident research would seem to stem from the lack of basic thinking about the causes of accidents in relation to the (individual differences) variable under scrutiny. Apparently, many researchers are only involved in thinking about, for example, the personality trait they are interested in, but stop dead before they reach the stage of exactly how this trait is manifested in behaviour which cause accidents, and which accidents they therefore should try to predict. Most just throw in whatever they get from their data files or self-reports, which results in a mixture of apples, oranges and melons. A select few researchers have, on the other hand, shown how variables are related to specific types of accidents, but more weakly to the overall category of traffic accidents, without explicitly considering why this is so.

In conclusion, it can be said that the available data regarding the effects of categorizing traffic accidents in individual differences research is scant, but what little there is points in the direction of this method having great possibilities. There are indeed great differences between sheep and wolves in this area of research, although few seem to have noticed this.

Chapter 2
The Validity of Self-reported Traffic Behaviour Data

... with traffic safety there is little doubt that progress depends on the development and use of instruments of proven reliability followed by the establishment of validity on a large scale. Grayson and Maycock, 1988, p. 235.

Introduction

For traffic safety research today, the dominant method for understanding driver behaviour is self-report. This is the case for independent variables, where many are hard to measure with any other method (attitudes, for example), but also for several parameters, which are objectively measurable; accidents, mileage etc. For example, in a typical individual differences study, a measure of personality will be acquired by self-report, and related to the similarly acquired number of collisions by each subject, and a significant association found.

In this chapter, the validity of some of these self-reported variables, notably crashes, will be investigated, and related to the methodological problem (or error) of using self-reported variables to predict other self-reports. It will be argued that this approach sometimes creates artificial associations and that this type of research therefore is non-informative at best, and often directly misleading. Also, the arguments in favour of this method by self-report researchers will be scrutinized.

Self-reports in traffic research are similar to pencil-and-paper tests used in many other disciplines. Some researchers administer their questionnaires using interviews, but in most instances, self-reports are mailed or handed out to potential participants in conjunction with some other measure being taken (on-line inventories are so far not very common). This type of investigation is very popular in driver safety research, as in many other fields of investigation, and probably for the same reasons, that is the instrument is in many ways very (even deceptively) easy to use, yields a lot of data, and is cheap in comparison to most other methods (for example, Hatakka, Keskinen, Katila and Laapotti, 1997). There is also the advantage that any question can be asked, any type of behaviour, attitude, thought or experience may be canvassed. The drawbacks of self-reports, however, tend not be acknowledged by the majority of traffic safety researchers, apart from some dutiful mentioning of the methodological limits of the study. The conclusions, however, do most often fail to bear out these statements of caution (for example, Iversen and Rundmo, 2004). Many researchers seem to assume that whatever is self-reported is true, and/or that whatever error is included is random. These assumptions are, as we shall see, untenable.

In many accident research areas, researchers have noted a strong memory effect; forgetting happens very rapidly. Also, various other effects make the validity of self-reports of incidents low to moderate (see Table 2.1 for references). It can also be noted that various biases of self-reporting have been found for any type of behaviour (Table 2.1). Actually, few researchers seem to have found any sizeable correlations between self-reported and actual behaviour. Among the exceptions are Barger et al. (2005), who reported correlations of .76 and .94 for hours worked and number of extended work shifts in a month for medical interns. For such a short time period, this could be expected, but, as will be shown, when the time period is longer (as in traffic safety research), correlations shrink rapidly. There is thus good reason to doubt that self-reports within driver research can be accurate.

Table 2.1   Studies on the validity of self-reports in two broad categories; injuries/accidents and behaviour/characteristics

Study	Type of incident
Gordon, Gulati and Wyon, 1962	Home injuries
Marsch and Kendrick, 2000	Home injuries
Warner, Schenker, Heinen and Fingerhut, 2005	Home injuries
Landen and Hendricks, 1995	Work injuries
Zwerling, Sprince, Wallace, Davis, Whitten and Heeringa, 1995	Work injuries
Klen and Ojanen, 1998	Work injuries
Valuri, Stevenson, Finch, Hamer and Elliot, 2005	Sports injuries
Langley, Cecchi and Williams, 1989	All accidents
Moshiro, Heuch, Åström, Setel and Kvåle, 2005	All accidents
Wallace and Vodanovich, 2003	Work accidents
Popham and Schmidt, 1981	Alcohol consumption
Evans, Hansen and Mittelmark, 1977	Smoking
Cummings, Nevitt and Kidd, 1988	Medical problems
Norrish, North, Kirkman and Jackson, 1994	Medical problems
Feldman, Cohen, Doyle, Skoner and Gwaltney, 1999	Medical problems
Harlow and Linet, 1989	Medical problems (review)
Wiggins, Schmidt-Nowara, Coultas and Samet, 1990	Snoring

Table 2.1   *Concluded*

Study	Type of incident
Engleman, Hirst and Douglas, 1997	Sleepiness
Barger et al., 2005	Amount of work
Ferrie, Kivimäki, Head, Shipley, Vahtera and Marmot, 2005	Absence from work
Robertson, Rivara, Ebel, Lymp and Christakis, 2005	Safety practices
Haapanen, Miilunpalo, Pasanen, Oja and Vuori, 1997	Medical problems

In general, in many other research areas, the discussion about the validity of self-reports and various biases and artefacts of this method is very much alive and present,[1] something that is not the case within traffic safety. Instead, a majority of papers published seem to be self-report studies, and the method is hardly ever questioned, criticised or validated.

The most important methodological problems concerning self-reports are their criterion-related validity and the possibility of common method variance (associations in data due to the method of acquiring them). If these issues have not been investigated, results cannot be trusted, because without these prerequisites it is not possible to interpret them. Any effect might be due to anything if it is not certain what is really measured.

The first problem is simply how well any given method actually measures what it is intended to measure. For self-reports, this could be the reported number of traffic accidents as a reflection of the real number. Common method variance is a term applied to associations within data sets that are not due to any real correspondence, but to a common way of measuring different variables. Self-reports are notorious for such effects, although this has hardly been noted within traffic safety research. One example of a simple mechanism that can affect self-reports is a scale response bias, where some respondents tend to use the extremes of the scales, whereas others tend to reply in the middle range, despite their actual behaviours/attitudes/experiences being the same. Within a sample, this will create an artefactual association between variables. This kind of self-report validity threat will be described further in a section designated to its effects within traffic safety research.

There are two main threats to the validity of self-reports about events; memory failures and various cognitive/social distortions. Unwillful erroneous reporting is apparently due to our fallible memories. Not only do we forget, but we also 'remember' things that have not happened, and in general change our perceptions of what has happened in accordance with later information. As has been shown

---

1   For example, Podsakoff and Organ, 1986; Feldman and Lynch, 1988; Johns, 1994; Barrick and Mount, 1996; Ferrando, 2008.

in eyewitness research, people are easy to influence (for example, Köhnken and Brockmann, 1987), which means that items as such in a questionnaire may also have an effect on the participants' replies. Also, various inherent biases seem to be at play whenever people report about accidents, like attributions (Burger, 1981) and stereotypes (Diges, 1988). However, while memory effects probably only add error into subsequent analyses, the other distortions may have effects on both independent and dependent variables, thus creating artefactual associations (common method variance effects). All self-reported variables therefore need to be validated, and the best method is comparison to other, more reliable sources, if such exist, thereby at least achieving a measure of the error involved in the data thus gathered.

In this chapter, the validity of self-reported traffic behaviour data will be reviewed and analysed. The reason for excluding attitudes and many other popular concepts is that these cannot really be validated in themselves, except by the use of some sort of additional hypothesis concerning their association with other variables. However, almost invariably, these other variables are also self-reported, un-validated and indeed impossible to validate. Here, the main focus will therefore be on variables that are objectively measurable, and where a fair amount of relevant reporting is available. This does not necessarily mean that the quality of the data is always good, but that there is enough information to discuss various possibilities and hypotheses.

Traffic Accidents

The importance of the accident variable within traffic research can hardly be overestimated. After a hundred years, it is still the most studied problem within the discipline, although other aspects of transportation have gathered momentum. It is therefore even more puzzling that so many researchers pay no heed to the validity of their main variable when they gather it by self-report. Actually, this problem is often not even discussed, despite there being in existence a number of papers that have, with various methods, tested the validity of self-reported collision involvement. In the sections below, a review of the literature concerning various aspects of self-reported accidents will study how people forget and distort incidents, how various sources agree, and how self-reported and recorded crashes differ in the results they yield.

Forgetting Crashes

Our analysis cannot resolve whether faulty recall affected the outcome of this study, but does suggest that self-reported accidents were not disproportionately associated with memory difficulties. Marcotte et al., 2006, p. 24.

People actually do forget that they have had vehicle crashes, even fairly severe ones. The evidence may not be plentiful (see Tables 2.2 and 2.4 for summaries), but the same type of results has been found with a number of methods, and agrees with findings in other research areas (see references in Table 2.1).

Starting with the test-retest agreement of self-reported accident involvement, Arthur (1991) found the correlation (presumably Pearson) to be .96 (N=111) for a period of only seven to ten days, which was replicated in another study with almost exactly the same result for two and three days (Arthur and Graziano, 1996). These values may seem impressive, but as subjects are most often asked to recall their accidents for periods of many years (af Wåhlberg, 2003a), a less than perfect agreement for a week's difference is a bit suspicious. Here, it is of some importance to notice that there is no reason to expect socially desirable responses[2] to have any effect on the correlation, unless one would like to claim that the respondents grew more susceptible to social pressure between tests (the second testing had a lower mean). Therefore, this difference is probably due solely to forgetting.

The conclusion about forgetting is supported by the results of studies by Maycock, Lockwood and Lester (1991), and Maycock and Lester (1995), who used the simple but ingenious method of asking people about their accidents per calendar year and comparing the means. As drivers' accident liabilities tend to decrease with age and experience, it could be expected that there would be a slight downward trend. However, the opposite was found; for each year, the number of self-reported accidents increased by 25–30 per cent,[3] indicating that people tend to forget this amount of incidents from previous years, or possibly, also tend to remember the dates of incidents incorrectly, placing them further back in time than is actually the case (as found by Cash and Moss, 1972). It would seem to be evident that people's memories are very fallible, even concerning such nasty incidents as traffic accidents; Cash and Moss (1972) found that 11.4 per cent of their sample of people who had been recorded as injured in a traffic accident less than a year ago did not even report the accident when asked, while other reported the crash but not the injury.

However, all traffic accidents are not that severe; some may be very slight and hardly worthy of notice. This difference also makes an impact on how people remember them. Chapman and Underwood (2000), using tape-recorded driving diaries, found that about 80 per cent of near-accidents were forgotten after two weeks,[4] with the more serious cases, and those where the driver was to blame,

---

2   A tendency to shape the answer according to some sort of perceived social norm.

3   A similar, somewhat weaker, tendency would also seem to be present in Davison (1985), but it is difficult to evaluate this result, as the details regarding the calculation of the means for accidents for different time periods were not reported.

4   However, it is important to point out a curiosity about the method employed; the drivers' recall and tape recordings were from different time periods. Thus, in contrast to almost all other studies on this topic discussed here, the criterion was not the same incidents, but presumed equal ones.

being less likely to disappear into oblivion. Also, the calculations of Maycock, Lockwood and Lester (1991) showed that when memory loss calculation was restricted to injury accidents, the rate of forgetting went down to 18 per cent.

Table 2.2   Studies that have compared agreement of self-reported traffic accidents/incidents with other sources, and reported per cent agreement in terms of the means of the sources

Study	Comparison	Value	N	Criterion	Time period	Comments
Chapman and Underwood, 2000	Decrease over time	80%	80	Near-accidents from tape-recorded driving diaries	2 weeks	Seriousness and culpability had effects, criterion and self-reports from different periods
Maycock and Lester, 1995	Increase per year	29%	8,617	Self-report for previous year	3 years	Reports for different years given at the same time
Maycock, Lockwood and Lester, 1991	Increase per year	30%	>13,000	Self-report for previous year	Variable, maximum 3 years	Reports for different years given at the same time

Note: Shown above are the type of comparison, size of the results, number of subjects, the criterion and time period used for comparison.

Furthermore, Maycock, Lockwood and Lester (1991) found that self-reported amount of driving in the dark and socio-economic group predicted the amount of forgetting. Although these results are somewhat hard to interpret, it should be very clear that forgetting of accidents is massive, and possibly has some systematic properties. So, a faulty remembrance of self-reported accidents is clearly a dire problem, which leads to under-reporting. However, there are reasons to believe that people under-report for other reasons than memory distortions, and that over-reporting of crashes may also exist.

Under- and Over-reporting of Crashes

> The result of drivers not reporting crashes would simply be to reduce the variability in crash involvement, thereby decreasing the strength of any potential relationships.
> Gras, Sullman, Cunill, Planes, Aymerich and Font-Mayolas, 2006, p. 136.

Fallible memories should lead to under-reporting of drivers' actual accident rate, while the main argument in favour of self-reported accidents is that these will yield a higher count than official registers, as found in about two thirds of samples (see Tables 2.3 and 2.4). However, this is not really an argument at all, but merely an observation of facts. A higher count does not automatically make these data more valid (which is the follow-up claim[5]), but may, actually, make them less so. First, it can be asked whether there are systematic biases of under-reporting, apart from the memory problem described?

One such finding was reported by McGuire (1976); drinking drivers tended to under-report accidents rather strongly (26 per cent). These drivers had a fairly high accident rate,[6] which make this finding similar to those for car drivers with many recorded accidents (Brown and Beardie, 1960; Harano, Peck and McBride, 1975; Owsley, Ball, Sloane, Roenker and Bruni, 1991). This could be interpreted as another memory effect; we have a hard time remembering each one of several repeated similar events, because they become amalgated.

However, in af Wåhlberg (2002a), about a quarter of the drivers reported fewer incidents than they actually had on record, and this result can hardly be due to these drivers forgetting more; those who reported less accidents than could be found in the archives not only had two and a half times as many accidents as the mean, they also reported fewer than the mean number. Furthermore, Lefeve, Billion and Cross (1956) found that 13 per cent of drivers interviewed claimed to have had no accidents at all during their driving careers, although state records accredited them with at least one. Such under-reporting cannot be explained by a memory amalgation effect. Instead, it could be suspected that some drivers have a tendency not to acknowledge their mishaps, that is, a social desirability effect, which will be discussed under the heading of common method variance.

In some situations it would seem to be fairly clear what the social expectation is, and that drivers do react to it when reporting collisions. Planek, Schupack and Fowler (1974) reported, regarding their evaluation of a defensive driving course:

> With reference to accidents, a discrepancy appears to exist between the per cent of reduction in the self-reported data and the state records for study group respondents. State record data indicate a before-after reduction in accidents of approximately 15 per cent less than shown in self reports. (p. 295)

---

5   For example, Sobel and Underhill, 1976; Nada-Raja, Langley, McGee, Williams, Begg and Reeder, 1997; Fischer, Barkley, Smallish and Fletcher, 2007; Barkley and Cox, 2007.

6   Almost twice the rate of the young drivers in McGuire (1973).

Table 2.3   Studies that have compared the numbers of self-reported traffic accidents with data from records of various types, and reported per cent agreement

Study	Agreement	Difference between sources	N	Criterion	Time period	Comments
Ball, Owsley, Sloane, Roenker and Bruni, 1993	27%	-	294	State records	5 years	Per cent people who reported the same number as the state source. The same set of data as in McGwin, Owsley and Ball (1998)
Barkley, Murphy, DuPaul and Bush, 2002	1/3	+	168	State records	Variable, lifetime	Agreement calculated from means in table 2
Barkley, Murphy and Kwasnik, 1996	26%	+	48	State records	Variable, lifetime?	Agreement calculated from means in tables 1 and 2
Begg, Langley and Williams, 1999	49%	?	1036	State records	3 years	Only injury producing crashes. The figure given was for percentage of total recorded crashes matched to self-reported (by details)
Brown and Berdie, 1960	21%	-	Probably about 80	State records	Variable	Selected group with many crashes on record
Cash and Moss, 1972	72.7%	Not applicable	590	State records	9-12 months	Interviews with people who had been in a reported accident, as driver or passenger
Dalziel and Job, 1997	90%	-	41	Insurance company records	2 years	Taxi drivers, matched accidents between sources
Findley, Smith, Hooper, Dineen and Suratt, 2000	33%	-	50	State records	2 years	It was not ascertained whether accidents were the same
Harano, Peck and McBride, 1975	59% Infinite	- +	196 (high) 231 (low)	State records	3 years	Contrasted groups. Reportable accidents only. Low group had no crashes on record but reported 0.11
Legree, Heffner, Psotka, Martin and Medsker, 2003	76.4%	+	551	Military records	5 years	Exactly how the agreement was calculated was not reported
Marottoli, Cooney and Tinetti, 1997	93%	+	358	State records	1 year	Old drivers

Table 2.3   *Concluded*

Study	%		Number	Source	Time period	Comments
McGuire, 1973	42%	+	500	State records	2 years	Matching of reportable crashes to state files. Only newly qualified drivers. Of the crashes said to have been reported, 62 per cent were in the files
McGuire, 1976	26%	-	500	State records	15 months	Drinking (alcoholic?) drivers
McGuire and Kersh, 1969	76% 32%	+ +	120 94	State records	Variable? Variable	Only reportable accidents
Owsley, Stalvey, Wells and Sloane, 1999	97.4%	-	182?	State records	5 years	Only incidents with police at the scene
Pelz and Schuman, 1968	90%	?	352	State records	1 year	–
Planek, Schupack and Fowler, 1974	87% 85% 87% 90%	+ + + +	2091 581 873 2091	State records	2 years	Trichotomized data: 0, 1, 2+
Ross, 1966	12.5%	+	36	State records	Variable, median 28 months	–
Schuster and Guilford, 1962	'about half'	+	306	State records	3 years	Self-reported accidents were twice as numerous as state recorded ones
Smith, 1976	46%	+	113	State records	3 years	Per cent of accidents found in both sources
Szlyk, Alexander, Severing and Fishman, 1992	<68%	+	31	State records	5 years	Maximum possible agreement calculated from raw data in paper
Szlyk, Severing, Fishman, Alexander and Viana, 1993	<83%	+	17	State records	5 years	Only crashes with police at the scene as criterion, maximum possible agreement (%) calculated from raw data in paper
Szlyk et al., 1995	Infinite	+	18	State records	5 years	No accidents were found in the records, while at least eight were reported
af Wåhlberg, 2002a	56.3%	-	150	Bus company records	3 years	Bus drivers

? denotes uncertain reporting.

Note: The agreement figures in per cent may not be exactly comparable between studies, as some have calculated their figure for number of accidents, and others for number of individuals with the same number between sources, while some have not reported the method used. A positive difference between sources mean higher numbers by self-report. Shown are the per cent agreement, which source had the higher mean, number of subjects, the criterion source and the time period used for comparison.

Table 2.4 Studies that have compared self-reported traffic accidents/incidents to other sources, and reported agreement in terms of an association measure

Study	Statistic	Value	Difference	N	Criterion	Time period	Comments
Arthur, 1991	Pearson ?	.96	Not applicable	111	Previous self-report	7–10 days	Test-retest
Arthur and Graziano, 1996	Pearson?	.96	Not applicable	2,27?	Previous self-report	2–3 days	Test-retest
Arthur et al., 2001	Pearson	.48	+	394	State records	6 years	–
Arthur et al., 2005	Pearson	.41	+	333	State records	2 years	–
Barkley, Murphy, Du Paul and Bush, 2002	Pearson	.41	-	?	State records	Variable?	Correlation reported in Fischer, Barkley, Smallish and Fletcher, 2007
Burns and Wilde, 1995	Pearson	.61	+	51	State records	Variable, mean 8 years	Taxi accidents
Dalziel and Job, 1997	Pearson	.90	-	41	Insurance company data	2 years	Taxi drivers, matched accidents, r estimated from other statistics
Edwards, Hahn and Fleishman, 1977	Pearson?	.25	Not reported	152	Probably state records	5 years	Taxi drivers
Marottoli, Cooney and Tinetti, 1997	Kappa	.40	+	358	State records	1 year	Old drivers
McGwin, Owsley and Ball, 1998	Kappa	.45	-	278	State records	5 years	Dichotomised data: accidents/no accidents. Only accidents where the police was present
Murray, Cuerden and Darby, 2005	Pearson	.399	Not applicable	16,000	Insurance records	2 years (records) 3 years (self-report)	Time periods only partially overlapping

Table 2.4   Concluded

Study	Statistic	Result	+/-	Subjects	Criterion	Time period	Comments
Owsley, Ball, Sloane, Roenker and Bruni, 1991	Pearson	.11	+	53	State records	5 years	Only crashes with police at the scene
Planek, Schupack and Fowler, 1974	Pearson	.38 .50 .45 .32	+ + + +	2,091 581 873 2,091	State records	2 years	Correlations calculated from tables 33-34 in study. Trichotomized data: 0, 1, 2+
Ross, 1966	Pearson	.21	+	36	State records	Variable, median 28 months	Correlation calculated from raw data
Stoohs, Guilleminault, Itoi and Dement, 1994	Pearson	.3	Not applicable	90	Truck company data	5 years	The coefficient is an underestimation of the actual association between sources, because private driving accidents were included in the self-reports
Szlyk, Fishman, Severing, Alexander and Viana, 1993	Spearman	.67 .52	+	8 17	State records	5 years	Only crashes with police at the scene as criterion
Szlyk, Seiple and Viana, 1995	Spearman?	.25	+	65	State records	5 years	–
af Wåhlberg, 2002a	Pearson	.38	-	132	Bus company data	3 years	Bus drivers
af Wåhlberg, Dorn and Kline, forthcoming	Pearson	-.17 -.05	Not applicable	300 115	State records Company records	1/10 years 2/variable	Self-reported accidents and state records for different periods for both samples

? denotes uncertain reporting.

Note: Shown are the statistic used, the result (association between sources), the difference of the means (positive meaning higher value for the self-reports), the number of subjects, the criterion and time period used for comparison.

The authors had four possible explanations for this phenomenon, one being study group respondents being 'less willing' to report accidents after the course.

There is thus evidence of systematic under-reporting. Whether this is due to forgetting or socially desirable responding or both is not really known, but the phenomenon as such would seem to be accepted by most researchers. The opposite (reporting crashes that have not occurred), however, does not seem to have been acknowledged (for example, Jerome, Habinski and Segal, 2006; Tronsmoen, 2008).

There seem to be two main ways in which over reporting of accidents could take place; first, erroneous memories for dates, which could put an event within the time boundaries for which the researcher has asked for, something which has hardly ever been researched (only Cash and Moss, 1972, has been found). Second, social desirability of a different kind than the commonly accepted type, where the respondent is assumed to try to look good to the researcher. When talking about socially desirable responding, all researchers seem to assume that accidents and risky driving are seen as undesirable (for example, Hatakka, Keskinen, Katila and Laapotti, 1997; Deffenbacher, Petrilli, Lynch, Oetting and Swaim, 2003; Lajunen and Summala, 2003; Fernandes, Job and Hatfield, 2007), and something which all people will confess to only with some reluctance. However, this is questionable indeed. In sub-cultures of young men, for example, motorcycle enthusiasts, it would seem very possible that crashing is seen very differently, probably as an unavoidable part of driving in very risky ways. Here, it should be possible to find people who even exaggerate their accident involvement. No research on this has been located.

However, the research would seem to indicate that we do not even need to go to such unusual populations to find over-reporting of accidents. McGwin, Owsley and Ball (1998) found that a few drivers (about five per cent) reported crashes that could not be found in official registers. Now, the significance of this is the very strict definition of accidents to be reported that was used by these researchers; incidents where the police had been called to the scene. Although this does not automatically mean that it will actually be reported, such circumstances are suspicious. Similar results were stated by McGuire (1973); only one third of accidents that were said to have been reported to the authorities could be identified (although this could to some part have been due to problems with accident details, see the section on Accident details and matching). Usually, this has been attributed to bad record keeping, but it could to some degree be due to social desirability, where a driver has admitted to an accident that is reportable under law, and then is asked whether it was reported to the authorities.

Also, it was found in af Wåhlberg (2002a) that close to 20 per cent of bus drivers reported a number of accidents, which could not be found in the company database, despite searches outside the time boundary in case of a mistaken remembrance of the date on behalf of the drivers (this happen to a fair per cent of drivers; Cash and Moss, 1972).

There is thus indeed evidence of both under- and over-self-reporting of traffic incidents, and having ascertained that, we need to turn to the overall association of self-reported accidents with other sources.

Self-Reports Versus Other Sources; Agreement

> ... past research has found self-report measures of crashes strongly correlate with objective measures of crashes ... Newnam, Griffin and Mason, 2008, p. 636.

When comparing self-reported accidents with recorded ones, it must first be acknowledged that, so far, no really proven dependable source of crash data is in existence. There are three or four possible types of archives with which self-reported accident data can be compared; state (police/hospital), insurance, fleet-based company, and military records. All of these have been used as sources for crash data in various studies, but only the first has been subject to extensive validation studies, with not very encouraging results. Still, when trying to validate self-reports, a comparison with records of some type is one of the few possible ways to go.

Most state sources are probably grossly inadequate concerning lesser incidents (for a review see Hauer and Hakkert, 1988). As noted, people generally report more accidents than can be found registered by police (Harris, 1990) and/or hospitals (see Tables 2.3 and 2.4). Concerning the validity of insurance company data, there seems to be little research available; Burg (1967; 1970) reported that an additional 51 per cent could be located in such a source as compared to the California Department of Motor Vehicles' archive. However, it is not quite certain if the accidents were matched against each other (see further the section on Accident details and matching about this problem). The third general category of accident records is archives held by fleet-based companies. Unfortunately, there seems to be no research available concerning their validity. Instead, one has to fall back on such arguments as the need for companies to keep accurate data, and their vehicle checking procedures. If vehicles are regularly inspected and investigations undertaken into un-reported damage, the validity is probably high (af Wåhlberg, 2002a; 2002b). Whether this is actually so in any specific case cannot be known, however, without checking company procedures. The situation would seem to be similar for military data, but there is even less information available regarding their data gathering procedures and record keeping. There is thus always the possibility of under-reporting finding its way into records, whatever method or register is used. However, many studies have reported on the congruence between self-reported accidents and other sources, and with the above shortcoming in mind, we might still get some general understanding of the real situation, especially if we compare the correlations between sources with a simulated example.

First, it should be noted that the accidents found in records have actually happened. No research into the validity of traffic accident records has ever found that any of the incidents described have in some way been fabricated, although they might be mislabelled, for example, concerning severity. This variance can therefore be seen as trustworthy. Furthermore, the tendency for records to contain mainly severe accidents is not such a strong drawback as it is usually made out to be, because the severe crashes are the most important ones to predict, and minor crashes are probably correlated with major ones within the individual anyway (although research about this is scarce, see Chapter 3). With these facts in mind, the reviewed research about agreement between sources can be analysed.

The literature on the agreement between self-reported accidents and other sources (mainly records of organizations) is summarized in Tables 2.2–2.4. It can be seen that, in general, the agreement is very low. The percentage agreement can of course be explained by the low validity of records for less severe incidents, but the correlations cannot. Here, the usual argument by self-report researchers is that this is due to the low validity of the archives. However, this is only an assumption, not something that has been proven, or even tested. On the contrary, it can be wondered how such low correlations can be found if there is a core of agreement about the severe accidents (that are more easily remembered). In af Wåhlberg (2002a), it was found that the correlation between records and self-reports of bus drivers' accidents was .38. Here, there was a core of agreement, but a quarter of drivers reported fewer collisions than could be found in the records, while almost as many reported more. If the former are excluded from this set of data, the correlation increases to .60 (N=103). This means that if all drivers reported as many or more crashes than the records, we could probably expect these variables to share at least a quarter of the variance. In a majority of the cases in Table 2.4, that is not what happened. Therefore, we can suspect that in most studies, there are negatively and positively differing self-report groups, as compared to the records. Correlations below .4 cannot be explained solely by drivers reporting more crashes than can be found in records.

This point can be illustrated by a simple simulation of results. Accident data for 500 drivers (m = 0.58, std = 0.84) was used, and defined as the archive variable. To test what would happen if drivers did report all the crashes that were found in the archive, plus a number of randomly distributed ones, crashes were added to the original ones in steps of ten per cent of the mean. Thereafter, the original and the expanded variable were correlated. The associations found were as followed; .96, .93, .91 and .89, for 10–40 per cent added accidents. This means that if there is a shared basis of variance, that is accidents, the association will remain very strong, even when fairly large numbers of incidents are randomly added to one of the variables. Furthermore, this means that it is improbable that the very weak correlations reported between archive and driver sources can be explained by an unsystematic difference in sheer numbers.

It can be noted that the size of the agreement between sources in the literature is weakly negatively associated with the time period used (Table 2.3: r=-.20, N=20; Table 2.4: r=-.16, N=18). The studies using a variable time frame were not included in these calculations, because the shared variance is probably artificially inflated,[7] which impacts upon the association between sources. Such a result could be expected from the memory problems described in the previous sections; the validity of archives tend not to be influenced by the time period used, while self-reports are, as longer periods make it harder to remember (correctly).

Regarding which source had more crashes, it is evident that self-reports are superior to state records in this respect, although there are a few exceptions from this rule. These instances are probably mainly due to restricted definitions of accidents, where only rather serious incidents were canvassed. For insurance and fleet-based data, what little results exist seem to indicate that they are superior to self-reports in terms of numbers.

A few researchers, apart from those in Tables 2.3 and 2.4, have reported good agreement between sources but given no figure for this association; Schuster (1968) compared state records with self-reports and reported no '... systematic bias in accuracy ...' (p. 18). Similarly, Larson and Merritt (1991) only said that '... comparison of subjects' responses with actual traffic records ... revealed an extremely high level of agreement.' (p. 39).[8] This result might be explained by the sample used; Larson and Merritt studied very young people (mean age 19.8 years) who must be presumed to have had very little driving and crash experience, thus making it easier for them to remember their mishaps. Also, Schuster and Guilford (1962), comparing self-reports with State Department files, found that for two-thirds of their subjects '... the information about violations and accidents from these sources agreed.' (p. 18). King and Clark (1962) mentioned that both self-reports and state records were used, and that 'Of the two accident indices, the higher one was assigned to each subject.' (p. 117), presumably meaning that there were cases when the state reported more incidents. West (1995), reported that, in his sample, the self-reported '... accident rates were similar to what would be expected from official statistics.' (p. 461).[9] If this means that they were similar in absolute numbers, it should mean that they were actually under-reported, as official statistics are. However, this was not the conclusion drawn by West.

For a slightly different type of self-report, one result can be mentioned; Alonso, Laguna and Segui-Gómez (2006) found that when comparing questionnaire and telephone interview reports about traffic injuries with ten months between them, 12 per cent first affirmed that they had been hurt and, at second interview, did not, while almost six per cent did the opposite.

---

7   Many researchers do not use a fixed time period for their accident variable, or any other exposure control (af Wåhlberg, 2003a, see Chapter 6), but different time periods for different drivers. This means that drivers with long experience will tend to have many accidents, and young drivers fewer. This is an effect that will be the same for both sources, thus creating an artefactual association that has nothing to do with the validity of either variable.

8   The traffic records mentioned were military (as were the subjects), and it may therefore be assumed that they were more valid than police or insurance company records.

9   The same argument was presented in West and Hall (1997).

Many researchers are in the habit of simply blaming state records for the weak correlations found between these and self-reports, while uncritically accepting the latter as valid. But as both under- and over-reporting of accidents exist, these must partly explain why the associations between self-reported and recorded accidents are low. The claim that the fault lies with the records is an assumption, and the opposite might as well be assumed, and as shown by the simulations presented here, the latter is more probable.

Accident Details and Matching of Reports to Records

> ... internalizers tend to give accounts of their accidents that show them to be culpable, while externalizers produce stories that depict themselves as blameless victims of outside factors. Such attribution factors ... could introduce artifacts or errors to investigations of accident causation that rely on victims' own statements of events. Foreman, Ellis and Beavan, 1983, p. 224.

Having established various shortcomings in the self-reporting of the numbers of traffic accidents would seem to be a fairly strong argument against this method. However, to really assess the validity of self-reports, one also needs to check how many recorded accidents are matched by the ones people self-report. Overall, there seems to be very little research into whether the accidents reported by drivers are actually the ones found in other sources, although it might be suspected that some of the studies discussed here (Tables 2.3 and 2.4) have done this, but not reported the results.

Given that the only way of knowing whether the reporting of an accident by a driver is actually matched by one on record is by asking about details (date, place, severity etc), it is very relevant first to ask whether drivers can actually remember such data correctly. Also, it is not uncommon to see research reports where the respondents are asked to provide precisely this circumstantial information[10] (although the reasons for asking are often obscure). Therefore, there should be some indications in the literature that drivers can actually report accident details in a valid way.

Dalziel and Job (1997) found the impressive figure of 90 per cent matching of the self-reported crashes of taxi-drivers (N=41) with insurance company records, and although they did not report a correlation coefficient, enough details were stated as to estimate the coefficient to be above .90. However, the circumstances were different from what is common in two aspects; a time period of two years for reporting was used, as compared to the more common three years (af Wåhlberg,

---

10   For example, Harano, Peck and McBride, 1975; Horswill and McKenna, 1999; Horwood and Fergusson, 2000; Parker, McDonald, Rabbitt and Sutcliffe, 2000; Sullman, Meadows and Pajo, 2002; Legree, Heffner, Psotka, Martin and Medsker, 2003; Fernandes, Job and Hatfield, 2007; Fergusson, Horwood and Boden, 2008.

2003a), thus making remembering easier, and it was also specifically noted on the questionnaire that answers would be checked, a method that probably reduces socially desirable responding. Here, these authors recommended the 'bogus pipeline' technique of Evans, Hansen and Mittlemark (1977), which means that the subjects are led to believe that their replies can and will be checked independently. However, this is unlikely to work for private drivers. Finally, and most importantly, Dalziel and Job did not state how the matching was undertaken, or how many corresponding details were needed to accept a match.

Begg, Langley and Williams (1999) reported figures for matching of details in themselves, running from 100 down to 76 per cent, for the variables vehicle, road user status, day of week, speed limit, number of years since crash, and time of day. However, these figures are probably strong overestimations of the real ones, due to the peculiar method used; self-reported crashes were matched to files, and the detail agreement was calculated only on those cases where matching was accomplished (the criterion for acceptance of a match was not reported). Similarly, Versteegh (2004) reported good agreement (>60 per cent) between self-reported and recorded time of crash, speed zone, environmental information and events of the crash.

In contrast to these fairly good matching figures, in af Wåhlberg (2002a) only about half of the self-reported accidents could be matched to reports from the archive, using a fairly lenient criterion,[11] consisting of such details as were regularly stated in the drivers' incident reports (af Wåhlberg, 2002b; 2004a). Also, Brorsson, Rydgren and Ifver (1993) found that about 45 per cent of drivers injured in car crashes stated that their recollection of the event was poor or very poor. Even more interesting was that this self-reported quality apparently was not associated with the accuracy with which the drivers reported the circumstances of the crash, as compared to police reports (the reported comparison was rather crude). One further indication about details of accidents can be found in Taylor, Chadwick and Johnson (1996), who reported that a small part (0.6 per cent) of their sample had given incompatible reports for number of crashes and details. Evidently, some drivers do not even understand what they are reporting about.

One of the problems involved in the evaluation of the research on validity of accident details and matching between sources is that the criterion for a match has usually not been reported, and a standard methodology for this is lacking. It can therefore be suspected that some results in this area are faulty, or at least not directly comparable to other.

For example, West (1997) calculated that the ratio of active to passive accidents (see Chapter 1 on taxonomies) in his samples, found them to be rather equal and claimed that if people had distorted their answers (a passive accident is probably seen as less blameworthy), there should have been a larger amount of passive than active accidents. This claim assumes that there should be an equal number of accident involvements fitting into these two categories. However, it could just as

---

11   Two or three agreements on the details, and no disagreement.

well be argued that there is a much larger number of active than passive accidents to start with, as two vehicles may very well strike each other 'actively', while the reverse is not true (see Chapter 4). Anyway, it is impossible to test what the 'real' ratio between active and passive crashes is/should be, because the definitions given are not, so far, explicit enough for such a test (see Chapter 1). This type of calculation can therefore not prove anything about people's distortion of accident details at all.

We are therefore left with virtually no evidence at all regarding the way people remember the details of their accidents, and we can only surmise that it is probably not better than the remembrance of the crashes in themselves, which, as we have seen, is not very good. It is therefore impossible to know whether the crashes drivers report are actually the same as recorded ones. In Chapter 4, the overall judgement of whether a driver was culpable for causing an accident (which a few researchers ask about), is considered.

Given the low agreement between sources for traffic accidents, it can be suspected that self-reported and recorded data will yield rather different results (that is your results will depend upon the method used), because they are so weakly correlated as to not being able to associate with the same variables to an appreciable degree. Whether this is what happens is the subject of the next section.

Predicting Accidents from Different Sources

> ... it may be that accidents with convictions, having been investigated by the police, are a more reliable criterion than accidents in general, most of which are self-reported. McKnight and Edwards, 1982, p. 190.

Given the proposed bias in self-reported traffic accident data versus other self-reports (see further the section on common method variance), it becomes very interesting to compare the results of studies using the same predictors but different dependent variables. If a bias exists, whether it be social desirability, congruent reporting or some other common method variance mechanism, the effect would be that a self-reported independent variable could predict self-reported accidents, but not recorded ones, or at least the effect would be much weaker. This difference is due to reporting tendencies creating shared variance where none exist in reality, or increasing an existing one.

However, this prediction is difficult to test, because self-reported and recorded crashes usually have very different means and variances, and any difference in association with independent variables in the expected direction might therefore be due to this difference instead (see Chapter 3 for an example of the paramount importance of the mean of accidents for the strength of associations). To be able to use this method, samples therefore need to be found where self-reported and recorded data have similar means. Only one study (af Wåhlberg, Dorn and Kline, forthcoming) seems to have undertaken this. There, it was found that the

Manchester Driver Behaviour Questionnaire could only predict self-reported collisions, despite higher means and validities for the recorded variables.

In another study where the difference between sources cannot be explained by a higher mean for self-reports, a correlation of .24 was found between self-reported bus accidents and self-reported car accidents, but while the first correlated .38 with archival bus accident data, there was no association at all for car accidents (af Wåhlberg, 2002a). This means that the drivers tended to think (or at least report) that they had similar accident rates while driving buses and cars, but it was doubtful whether this was really the case.

The use of self-reported accidents as outcome variable has mainly been argued for in a negative way, being based upon the unreliability of state records (for example, Fischer, Barkley, Smallish and Fletcher, 2007), but, as pointed out, the most important information (severe accidents) actually has a very good coverage in these sources (for a review, see James, 1991), and the added variance of self-reports should only serve to make it easier for researchers to find significant effects, in a statistical sense. The relative strength of the predictor variables should be the same between sources, however, if self-reported collisions have a reasonable validity. This means that if a certain self-report study finds risk-taking to be the most important predictor of accidents, out of a set of predictors, then a study using recorded crashes as outcome should find the same thing, if they are using the same predictor data. The only difference would be that the effect would be weaker, due to lower variance. If self-reported accidents tend to be predicted by other variables than recorded ones, there exist a problem. What use is it to predict small dents if the large ones cannot be predicted? And why would this happen? It can be suspected that such an effect would be due to that some self-reported predictors are prone to common method variance effects, while others are not.

A few studies have computed associations for self-reported and recorded accidents in parallel, using state-recorded data. These cannot be evaluated by comparing effect sizes directly, as the means of self-reported collisions usually are higher. However, it is possible to use the secondary correlation method, described in the introduction, because it could be suspected that recorded and self-reported collisions would yield different patterns of association with various predictors, as some of them would be spuriously inflated, due to common method variance (see below). What few such results that were possible to calculate can be seen in Table 2.5. It can be concluded that these two sources seem to yield rather different results. This means that a variable that is a relatively strong predictor of traffic safety as measured by self-reported incidents is unlikely to be so by recorded ones. It is important to point out here that the absolute size of the effects are not the issue here, but rather the relative standing of different variables as accident predictors in different sources.

In the two examples in Table 2.5, it is evident that self-reported and recorded data yield rather different association patterns, as these share less than half their variance. Such results can also be detected in many studies that have used too few variables for a secondary correlation analysis.

Table 2.5   Correlations between effect sizes for self-report and state-recorded accidents with different predictors in various studies

Study	Predictor variables	Correlation between effects	N	Original N	Original statistic	Mean effect self-reported accidents	Mean effect state-recorded accidents	Comments
Arthur et al., 2005	Big Five personality and driving speed	.61	12	333	r	.080	.065	Data taken from table 3 in study
McGwin, Owsley and Ball, 1998	Vision	.68	12	278	Percent	–	–	Data taken from table 2 in study. Different variables significant for the dependent variables

Note: Shown are the predictors, the N of the correlation (number of predictors in study), the N used to calculate the original effects, the original statistic, and the mean effects for each criterion source.

A few other results in this vein may be mentioned; Burns and Wilde (1995) reported that; '… the pattern of the results did not change when driving record collisions were substituted by self-report collisions.' (p. 273). However, it was not reported how the comparison between results had been undertaken, or any value regarding their similarity. Given the correlation of .61 between accidents for these two sources, it is hard to understand how similar results could be possible. Owsley, Ball, Sloane, Roenker and Bruni (1991), found many significant correlations between self-reported crashes and their independent variables, despite their r=.11 with state-recorded accidents, which also correlated significantly with the predictors, but presumably with other parameters.

It can therefore be concluded that the evidence suggests that self-reported and recorded crashes yield different results from the same predictors, as could be expected from the mechanism of common method variance. This difference cannot be explained by the characteristics of archives, neither in terms of statistical variance nor reporting tendencies.

Discussion and Conclusions; Accidents

… questionnaires and interviews must be considered doubtful sources of information in relation to road traffic accidents. Lings, 2001, p. 438.

Summing up the chapter so far, it would seem that the, somewhat limited, research on the validity of self-reported collisions indicate:

1. strong forgetting;
2. under-reporting by those with many crashes;
3. possible over-reporting by some sub-groups;
4. low agreement with other sources;
5. uncertainty about the correctness of reported details;
6. different predictive patterns as compared to other sources.

A few other problems can also be suspected, although there is not enough evidence available today for any certain claims. It can be presumed that self-reports actually miss vital data, due to the self-selection of participants, as Planek, Schupack and Fowler (1974) reported that drivers on a defensive driving course who did not respond to a request of filling out a questionnaire had 20 per cent more accidents on record than the responders. Similar effects were found by Donovan, Queisser, Salzberg and Umlauf (1985) and Ferdun, Peck and Coppin (1967), although the latter was not significant, as was probably not the slight (4.7 per cent) difference found by Stutts, Stewart and Martell (1998). Chipman (1982) reported a different type of selection bias; drivers with many demerit points on their licence were harder to reach, apparently because they were rarely at home or had moved and left no forwarding address. However, the results were not found to be influenced by this association. Bygren (1974) found that non-responders had higher mileages than responders.

On the other hand, Humphriss (1987) found that drivers who had not reported their accident to the company they worked for had very similar scores on a vision test as compared to those who had reported their mishaps (both were different from a no-accident group),[12] and Guibert, Duarte-Franco, Ciampi, Potvin, Loiselle and Philibert (1998) found no response bias, comparing for police-recorded crashes.

So, the drivers that would add most variance may opt out of surveys, while a state record-based study would not have this problem, unless the predictors were gathered by self-report. However, as the majority of traffic safety research at the individual level use self-reports as predictors, this point would seem to be moot anyway, apart from being an incitement for objective measuring.

It might also be argued that the, sometimes strong, associations found within self-reported data would need to be ludicrously strong in reality, given the various biases discussed so far. For this reason too, it may be suspected that many associations found in studies using self-reported data are spurious, or at least overestimated, due to various cognitive mechanisms, as will be described in this chapter in the section about common method variance.

---

12   This study is a bit peculiar, as it is the only one noted where drivers with accidents were older (mean 40 versus 36 years) than those without collisions.

It should be noted that the results reviewed here, and the conclusions drawn, only apply to self-reported accidents when they are used with the individual as the unit of analysis. When group means are used, the different biases would seem to cancel each other out, and the results come out pretty much the same as for state records, at least according to what little research there is in this vein (for example, Roberts, Vingilis, Wilk and Seeley, 2008).

Having established the many problems involved in self-reported traffic accident data, we turn to a few other popular self-report variables that are often used in conjunction with crashes, as independent or dependent measures. As will be shown, these variables share many of the problematic features of crashes, when it comes to self-reporting them.

Exposure

> The main study was preceded by a pilot study to test the questionnaire and establish the accuracy of trip recall ... This phase ... established the feasible recall period as three days. Toomath and White, 1982, p. 407.

Exposure is an important variable in traffic research; for risk calculations and prediction of accident involvement, it must somehow be held constant. After all, 'differential accident liability' and similar expressions will not have much of a meaning if the environment/exposure is allowed to have an influence (see also the discussion about the non-use of exposure control in Chapter 6). Regarding the validity of self-reported driving exposure, again the research is rather sparse.

However, White (1976) compared self-reported mileage with actual readings on the odometer (obtained from a State car inspection program), and found that high mileages of vehicles (not drivers) were underestimated and low mileages overestimated. The mean could therefore be used with a fair degree of certainty, but association measures would be unreliable. Similarly, Huebner, Porter and Marshall (2006) found low correspondence between self-reported and recorded mileage (r about .58[13]), although the difference between the means was less than ten percent. Only Chipman (1982) seems to have reported a sizeable correlation (.86) with the odometer method. This was for only one week's driving, which probably explains the much higher value found.

While White (1976) and Huebner et al. found under-estimation of high mileages and vice versa, Staplin, Gish and Joyce (2008), reported the opposite, and also referred to two other papers. It can therefore be suspected that the validity of self-reported mileage data is influenced by how the actual question is phrased; Owsley, Ball, Sloane, Roenker and Bruni (1991) used several different ways of eliciting the same information, and found only moderate agreement between them (r's from .22 to .68). In another study by the same research group, the conclusion

---

13   This value was estimated from a scatter plot in the paper.

from a similar undertaking was that the results were too unreliable to be used for exposure corrections (Ball, Owsley, Sloane, Roenker and Bruni, 1993). Guibert, Duarte-Franco, Ciampi, Potvin, Loiselle and Philibert (1998), on the other hand, reported a correlation of .78 between two different reporting methods for weekly mileage.

Apparently, people seem to have trouble calculating averages for different time periods that are mathematically equivalent, thus yielding differing answers. Such results have been reported by Staplin, Gish and Joyce (2008), and McGuire and Kersh (1969). For the latter study, the correlation between self-reported overall and a sum computed from reported driving habits correlated .74. However, it was also reported that almost a third of the respondents had no idea what so ever about their mileage.

However, even when questions are identical, responses probably differ between modes of reporting. Alonso, Laguna and Segui-Gómez (2005) thus found that the intraclass correlation between mileage reports from questionnaires and interviews (separated by about ten months) was .64. This result could therefore be explained both by separation in time and different methods. It should be noted that the time period for which answers were elicited was the same for both modes.

On the more positive side, Joly et al. (1993) found that truck drivers could estimate their mileage for their last day worked with rather good precision (.82, Pearson correlation), while time spent driving was not equally well reported. The opposite was found for bus drivers (but no correlation reported). Also, car drivers keeping a driving diary yielded the impressive figure of .90 between this value and a formerly given estimation of a previous weeks' driving. Joly et al. also listed several methods for cross-validating self-reported mileage, for example, self-reported income for professional drivers.

As the only really high validity values for mileage have been found with extremely short reporting periods (as compared to what is common within traffic research, af Wåhlberg, 2003a), the conclusion for this variable is similar to that of the crash variable; it is too unreliable to be used in individual differences research.

Seatbelts

The wearing of seatbelts and the methodology for measuring this has received some attention from researchers. Thus, Hunter, Stewart, Stutts and Rodgman (1993) found that 13 per cent of the drivers who had reported that they always wear a seatbelt (N=2,507) in fact had been observed without it at the time of the questionnaire being distributed. Somewhat curiously, the opposite was observed for six per cent of those who said they never wore a belt, replicating the results of Hunter, Stutts, Stewart and Rodgman (1990). These discrepancies occurred although the questionnaire also had a question whether a belt was worn at the time of the distribution. Similar discrepancies were found by Streff and Wagenaar

(1989), Stulginskas, Verreault and Pless (1985), and Webb, Bowman and Sanson-Fisher (1988), who also found a slight under-reporting, while Fhanér and Hane (1973) reported a correlation of .73 between observed and reported seat belt wearing. Markey, Buttress and Harland (1998) found that at least 15 per cent of observed rear-seat non-wearers reported always using a belt in the rear seat. Finally, Robertson (1992) compared reports and observation percentages for several states and found consistent and substantial over-reporting, as did Dee (1998).

Also, Parada, Cohn, Gonzalez, Byrd and Cortes (2001) found systematic differences in reporting; those who tended to use the seatbelt less over-reported more (in this case it was ethnic groups), while Stulginskas, Verreault and Pless (1985) found associations with various socio-demographic variables such as parents' educational level.

So, the available evidence for self-reports about seatbelts is also in agreement with the previous sections; such data is unreliable and spurious effects are very possible, although this does not seem to have been investigated. It could, however, be predicted that in countries like Sweden, where there is a strong positive social pressure regarding the use of seatbelts, social desirability would be evident if investigated. As with other self-report variables, the unreliability of seat-belt reports does not seem to have registered with researchers (for example, Blows, Ameratunga, Ivers, Lo and Norton, 2005).

Speed

> Social scientists have known for many years that there may be notable discrepancies, and even an inverse relationship, between reported and actual behavior. Popham and Schmidt, 1981, p. 355.

Many traffic safety researchers are in the habit of using proxy safety variables, like violations and near accidents, instead of the real thing; crashes. Here, speed is one of the most commonly used, due the widespread acceptance of the importance of this variable. This acceptance is probably mainly due to the behaviour being easy to understand and measure,[14] and it will be argued in Chapter 7 that even when validly measured, this variable is not a valid proxy for collisions in individual differences research. However, first speed will be scrutinized regarding its validity when self-reported.

Using in-vehicle observers, West, French, Kemp and Elander (1993)[15] found correlations of .55–.65 between self-reports and observed speed, for

---

14   It could also be due a logical error, where the findings of accident investigations are applied to individual differences research without considering the base rate of the phenomenon, which will be described in Chapter 7.

15   This data has also been published as part of the report West, Elander and French (1992).

various measures. Although this still leaves more than half of the variance unaccounted for, these are among the best results ever found, probably due to the more continuous nature of the measurements. In contrast, Corbett (2001) found associations of .09-.17 (Kendall's Tau c[16]) between various self-reported speeds and as measured by speed cameras (on the same road) in her samples. This extremely low correspondence was due to speed being mainly under-reported, but also somewhat over-reported by about 16 per cent of drivers. Haglund and Åberg (2000) found that drivers could self-report with a correlation of .58 to measured speed, when there was a distance of one to two kilometres between measurements (that is a few minutes time between actual driving and reporting). This correlation was considered strong. Similarly, Vogel and Rothengatter (1984, as reported by Lajunen and Summala, 2003), found a correlation of .56 between self-reported and actual speed. Unfortunately, the exact circumstances of the comparison were not reported.

Here, it is instructive to note that Åberg, Larsen, Glad and Beilinsson (1997) found correlations of .36 in two samples of drivers (with a methodology similar to that in Haglund and Åberg, 2000) between observed and reported speed for the same stretch of road. However, the latter was much more strongly associated with reported normal speed (r=.61, .70). This is an example of common method variance.

Harré (2003), in an experimental field study, found extremely large and systematic differences between drivers' self-estimated speed and their measured velocity, due to the presence of children by the roadside. If no one was present on the curb, mean values were very similar. However, when two children were playing beside the road, mean estimated speed went down 30 per cent, while the real difference was only 2 per cent. If the two children were waiting to cross the road, a further reduction resulted, still with the estimated speed changing more. The estimates were made by the drivers 500 meters down the road. Similarly, Lourens, van der Molen and Oude Egberink (1991) found that drivers' responses on questionnaires and in traffic to encounters with children correlated .06 for bicyclists, and .57 for pedestrians. However, it was not explicitly described how the situations had been observed.

As described earlier, Versteegh (2004) found fair agreements between drivers' reports of several of their accident details and records, but this did not include speed. As compared to the results from the investigation of the crash, only 3.8 per cent were correct. The time from incident to report was 3.5 months (mean). Unfortunately, it was not stated what cut-off difference between investigation estimate and self-reported speed was used to determine 'correct'. The mean under-estimate was 13.6 km/h.

Last, two studies have been found that have used self-reported and recorded speed in parallel as outcome measures. Paris and Van den Broucke (2008),

---

16   The reason for using this statistic was probably that the self-reports were in six categories instead of continuous.

compared how well these speed variables could be predicted by attitudes to speeding. As expected, self-reports predicted self-reports, but not data from an on-board computer. It was not described whether any other drivers had access to the cars used by the subjects, a possible source of error that could explain the results.

A similar effect was found by Åberg and Wallén Warner (2008), where self-reported speeding was almost twice as strongly associated with various other self-reports as compared to logged speeding. However, as the logged speeding variable was not exclusively based upon the same drivers as the self-reports (other had been driving the same car), these results are hard to evaluate.

Again, the available research show self-reports to be inaccurate and under the influence of systematic biases. No one has reported results where even half the variance between self-reports and objective measurements was shared, and as the standard argument from self-report researchers trying to explain why crashes do not correlate well between sources (the criterion source is unreliable) is not really applicable for this variable, this is indeed alarming. It could of course be argued that overall judgements (like 'I always speed') are more dependable than spot measurements (what speed were you doing before that bend?), but until results show this, it remains a hypothesis, and it is really the responsibility of those who make such claims to prove them. Unfortunately, it would seem that it is often the researchers who distrust the self-report measures who test them, instead of the proponents, where the burden of proof should lie.

Citations/Violations

> The need to validate self-reported accidents and violations against official files is recognized ... It is expected, however, that more traffic offences may be reported in a confidential interview than are verifiable in the files. Schuman, Pelz, Ehrlich and Selzer, 1967, p. 106.

As with speed, citations[17] for traffic infringements is a very popular proxy outcome variable in traffic safety research. Here, only the validity of self-reported citations as a measure of the true number will be discussed, while the validity of this variable as an individual differences in safety outcome variable is treated in Chapter 7. Also, some methodological consequences of a peculiar similarity between these self-reports and those for collisions will be noted. First, the literature on the validation of self-reported citations will be described.

A few comparisons of self-reported and state-recorded citations have been located, with results that can be seen in Table 2.6. The associations would seem to be of the same strength as those for accidents, while over-reporting of these

---

17   'Violation' is often the term used for traffic tickets, but unfortunately, this word is also used for traffic infringements that are illegal, but where the driver has not been caught. In this section, only the first meaning is intended.

Table 2.6   The associations and differences between self-reported and state records for traffic citations

Study	Correlation between sources	Difference	N	Type of citations	Time period	Comments
Arthur et al., 2001	.59	305%	394	Moving violations	6 years	–
Arthur et al., 2005	.22	533%	333	Moving violations	2 years	–
Brown and Berdie, 1960	–	-33%	993	All but parking violations	4–6 years	–
Barkley, Murphy, DuPaul and Bush, 2002	.39	129% (ADHD) 129% (control)	132 168	Total tickets	Variable	Uncertain reporting about N, possibly unequal numbers for the means. Calculated from table 2. Correlation reported in Fischer, Barkley, Smallish and Fletcher, 2007
Barkley, Murphy and Kwasnik, 1996	–	73% (ADHD) 364% (control)	25 23	Traffic violations	Variable?	–
Corbett and Simon, 1992	–	235%	457	All	Three years	–
Fischer, Barkley, Smallish and Fletcher, 2007	.52	185% (ADHD) 289% (control)	147 74	All	Variable	–
McGuire, 1973	–	31%	500	Moving violations	2 years	–
McGuire, 1976	–	-36%	500	All	15 months	–
McGuire and Kersh, 1969	–	-2%	122	All?	Not reported	Data taken from table 5 in study.
Ross, 1966	(.09)	50%	36	All	Variable, median 28 months	Correlation coefficient calculated as a rough estimate, based upon raw data
Smith, 1976	–	254%	113	Convictions	3 years	Nine of 24 claimed convictions were not the same as the recorded ones. Only convictions for serious offences were usually recorded.

? denotes uncertain reporting.

Note: Shown are the correlations between numbers in these sources, the per cent difference (positive for higher self-reported numbers, always in per cent of the state records), N of study (that is, subjects, not number of citations), the types of citations included, and time period for comparison. Moving violations exclude those for parking and equipment offences.

instances seem to be very strong. No research concerning the reliability over time of reports of citations has been found.

When self-reported citations have been compared with state records, the ensuing moderate correlations have, if commented upon, been explained by the same kind of 'logic' as presented for accidents; the error is assumed to lie with the records, because more violations are reported by the subjects. However, this result is very strange, because the sequence of happenings for citations and crashes are very dissimilar. For the latter, it is a fact that most incidents do not come to the attention of the authorities, so a higher mean for self-reports is a natural outcome, while the low correlations between sources is not. Citations, however, are the result of interactions between drivers and police officers, and there should be no under-reporting in the records, unless the data have been lost or purged, because the very definition of a citation is that a driver behaviour has come to the attention of the authorities.

Yet another explanation would seem to be that respondents are reporting citations that they have had in some other state than the one used for comparison. These possibilities were pointed out by Barkley, Murphy and Kwasnik (1996), regarding drivers with Attention-Deficit/Hyperactivity Disorder (ADHD), concluding that it was also possible that '... the subjects may not always be completely accurate or truthful in their self-reports ...' (p. 1093). As there was massive over-reporting of violations in their study, it would seem like the most probably explanation is that the subjects were reporting violations that had never happened. The researchers themselves, however, claimed the opposite, believing that the ADHD subjects may under-report adverse driving outcomes (for example, Fischer, Barkley, Smallish and Fletcher, 2007) in a similar fashion to their under-reporting of ADHD symptoms.

The possible over-reporting effect for ADHD subjects will be further studied under the heading of common method variance. First, however, we will take a look at a very popular driver behaviour inventory, because it can be suspected that there are validity problems for such measures too.

DBQ: Errors and Violations

> I do not believe anyone would argue with the use of self-report measures when the theory or construct involved is attitudinal or perceptual, but when the reason for the use of self-report measures is one of convenience only, most investigators begin to view their use critically and to evaluate the degree to which some form of method bias might constitute an alternative explanation of the investigator's results. Schmitt, 1994, p. 393.

As noted in the introduction of this section, the main advantage of self-reports is that it is possible to ask any question, and get an answer. However, this is also the weak point; it is obviously very tempting for researchers to ask all these hard questions, but doing so does not guarantee answers that are fair reflections of reality. Unfortunately, many seem to have forgotten this.

As should be apparent from the preceding sections, most studies of traffic-related behaviour variables that have compared self-reports and other sources have shown the former to be biased in some way. But these variables are a minority in the meaning that the majority of self-reported traffic behaviour scales (which are often a mixture of several inter-related behaviours) have not been validated against more reliable sources for the same behaviour. The strange thing is that, with one exception, this goes even for the most popular driver questionnaire in the world; the Manchester Driver Behaviour Questionnaire (DBQ).[18] This inventory measures what is called aberrant driving behaviours; intentional violations of safe driving practices, and various driving errors that can be the result of lack of attention, erroneous decisions and/or unintended actions. Although many of the items are about intentions versus outcomes (for example, mistakenly put on the wipers instead of the lights) that cannot really be compared against an objective source, there are some that could be validated by observation.[19]

Despite the great popularity of the DBQ, there would seem to exist only one validation study of this questionnaire in terms of comparing reported with actual behaviour; Rimmö, 1999.[20] In this study, a small group of subjects drove a pre-arranged route of 50 km, with two observers in the car. Afterwards, the subjects completed the DBQ for aberrations made during this drive, which was compared with the same type of data from the observers' notes. The highest association found was .58 between one observer's ratings and the subjects' replies. Although the analysis was not reported in full detail, it must be concluded that the correlations computed were not at the level of single items, but factors, meaning a substantial aggregation of the data per individual. Also, the subjects' reported aberrations during the test were not significantly correlated with their responses for their normal driving. Rimmö himself argued that the driving undertaken was not representative for the drivers' everyday driving, and that this difference explained the latter lack of correspondence. However, it is hard to see how it could explain the low association between observers' results and self-ratings.

---

18   For example, Blockey and Hartley, 1995; Parker, Reason, Manstead and Stradling, 1995; Parker, West, Stradling and Manstead, 1995; Lawton, Parker, Stradling and Manstead, 1997a; Stradling, Parker, Lajunen, Meadows and Xie, 1998; Åberg and Rimmö, 1998; Dobson, Brown, Ball, Powers and McFadden, 1999; Rimmö and Åberg, 1999; Parker, McDonald, Rabbitt and Sutcliffe, 2000; Kontogiannis, Kossiavelou and Marmaras, 2002.

19   For example, 'On turning left nearly hit a cyclist who has come up on your inside'.

20   It is somewhat confusing that Rimmö (1999) seems to claim that West, French, Kemp and Elander (1993) and Groeger and Grande (1996) made validation studies on the DBQ. The first did not use DBQ but somewhat similar scales, while the second did not compare the DBQ results with aberrant driving, but with 'assessed performance'. Lajunen and Summala (2003) may also give the unwary reader the impression that the Groeger and Grande study tested the DBQ factors against observations of the same behaviours, but a close reading does indicate that they were aware that it did not.

It should also be noted that it is not as claimed by Parker, West, Stradling and Manstead (1995); 'Rolls, Hall, Ingham, and McDonald (1991) have reported a high level of correspondence between self-reported driving behaviour as measured by the DBQ and observed driving behaviour on a 40 km test route.' (p. 575). Very similarly, Sullman, Meadows and Pajo (2002) wrote; 'Rolls, Hall, Ingham, and McDonald (1991) reported a high level of correspondence between self-reported driving behaviour (measured by the DBQ) and actual driving behaviour on a 40 km test route.' (p. 230). This claim was repeated by McKenna, Horswill and Alexander (2006), while Gras, Sullman, Cunill, Planes, Aymerich and Font-Mayolas (2006) cited the Rolls et al. report as proof of a significant relation between self-reported and actual driving behaviour.

However, although Rolls et al. used eleven items, which were very similar to DBQ questions, they did not report any comparison between these and observed driving. Actually, their only comment in this vein was considering the variables ability, safety, anticipation, concentration, observation and car control; 'Correlations between the observers' assessments and the drivers' assessments were mostly only marginally significant, which confirmed that the drivers' perceptions of their abilities were inconsistently different from those of the observers.' (p. 54). With 411 subjects, the number used by Rolls et al., a Pearson correlation of .10 is significant at p<.05, indicating that the correlations, which were not reported, must have been very weak indeed.

It should also be remembered that the testing undertaken by Rimmö (1999) and Rolls et al. were both very dissimilar to how self-reported traffic behaviours are usually canvassed; by questions pertaining to rather long time periods (these are most often not stated exactly, but terms as 'usually' are common). In contrast, the data in these studies was instead gathered for very short periods, actually in relation to a specific test drive. Therefore, even if the correlations between self-reports and behaviours had been high, it would not have been of much consequence for the DBQ or any other driver behaviour inventory. As we have seen for other variables, people may be able to report fairly accurately about their behaviour for short time periods adjacent to the reporting event, but this ability rapidly detoriates as the time period is extended.

Furthermore, the sweeping statement of Parker, West, Stradling and Manstead quoted above, and the echoes from other researchers, could not possibly be true for the whole DBQ, even if their reference had undertaken something of the sort claimed. As noted, many of the items are about differences between intended and executed behaviour, a difference that simply cannot be observed.

It can therefore be concluded that the world's most popular driver questionnaire is unvalidated and, given the results for other self-reported behaviours, probably a poor estimate of the actual 'aberrant' driving. What is usually given as arguments for its validity is instead that its factor structure has been replicated many times, and that the violation factor predicts accident involvement. Both of these statements are dubious, as evidence would seem to show that the factor structure is rather dissimilar between studies, and that the association with (self-reported) crashes is due to common method variance (af Wåhlberg, submitted; af Wåhlberg, Dorn and Kline, forthcoming). We therefore turn to this methodological concept, which may explain a large number of results within traffic safety research, not only those of the DBQ.

Common Method Variance in Self-Reported Data

> Furthermore, the use of self-reports to measure both personality and accidents limits the study because any correlation may be a response artifact. Arthur and Graziano, 1996, p. 599.

The term common method variance (CMV) is an umbrella name for various effects between variables that occur due to the data source being the same. In reality, it is a problem that is almost solely restricted to self-reports.[21] The mechanism might be as simple as a tendency of subjects to respond to scales in a particular fashion (that is one person uses the extremes, another does not). Such tendencies may create associations in the data where none exist in reality.

One argument commonly proposed in favour of self-reported accidents by many researchers using them is that they have found significant associations with other variables, despite the shortcomings (dubious validity) of their parameters. Turning sin into virtue, they claim that if correlations can be found despite a large error component, the true association must be much stronger (for example, Parker, McDonald, Rabbitt and Sutcliffe, 2000). This is true if the independent variables come from another source than the dependent, preferably an objective one (as, for example, in Marcotte et al., 2006). However, the dependent variables are most often also self-reported, and there is thus good reason to suspect that some sort of common method variance (Campbell and Fiske, 1959) cause inflated, if not totally spurious, effects in these studies. Several such mechanisms have been proposed and tested in other research areas (for example, Podsakoff and Organ, 1986; Podsakoff, Mackenzie, Lee and Podsakoff, 2003); acquiescence, consistency, likeability, similarity, negative affect, carelessness and social desirability (Schmitt, 1994; see also Feldman and Lynch, 1988). Only the last seem to have been imported into traffic safety research, and the present discussion will therefore mainly centre on this mechanism; the wish to present oneself in a favourable way.

---

21   One example of a common method variance effect outside the realm of self-reports is accidents and violations in state records. As is shown in Chapter 7, these two variables might correlate fairly, and yield very similar effects for other variables, but this may at times be due to tickets being issued for crashes. Unfortunately, it has seldom been stated in the published research whether such violations have been included or removed from the data.

Socially Desirable Responding

Social desirability is defined as a tendency to present oneself in a favourable way, that is, more or less consciously lying with an intention of making people think that you are a nice person. Such concepts are traditionally measured within psychology with the use of so-called lie scales, that is, items that are constructed to measure a propensity not to tell the truth. The basic logic is that people who want to show themselves in a positive way will tend to respond positively to very exaggerated statements about honesty and similar concepts, for example, 'I have never stolen anything, not even a hairpin'. As such saintly behaviour is very unlikely in humans, these claims are taken as evidence that the respondent is not telling the truth. It should be noted, however, that this measurement method only canvasses the tendency for socially desirable responding in terms of some sort of very general, politically correct social norm, which the respondents perceive that they and the researcher (or anyone asking the questions) probably embrace. If respondents adhere to some different norm, they might try to impress others with exaggerated conformance to these rules. The latter proposition is hypothetical, though, because no research seems to have investigated this kind of alternative social desirability.

If social desirability affects both dependent and independent variables, spurious associations between these will result. Several questions can therefore be put to the literature:

1. Is there an association between self-reported collisions and social desirability?
2. If 1) is true, is this due to a reporting tendency, or a real effect, that is are drivers high on social desirability actually better drivers?
3. Are self-reported independent variables (often driver inventories) affected by social desirability?

For common method variance to be a problem in accident prediction studies, 1) and 3) must be true, and 2) not true. It should be pointed out that the available research is really only about the common conception of social desirability in relation to the researcher, not towards any other group with different concepts about what is socially desirable. This, however, may be enough to create spurious effects.

Social Desirability and Accidents

> Use of self-reported accidents raises the possibility that a 'dishonesty' bias might underlie associations found with attitude or personality measures. West and Hall, 1997, p. 255.

First, we need to review the studies on social desirability versus accidents and driver inventory responses. Only Lajunen, Corry, Summala and Hartley (1997; 1998) seem to have developed and used (Lajunen and Summala, 2003) a driver social desirability scale (DSDS) for traffic research, and this has not aroused much interest (for exceptions, see Caird and Kline, 2004; Sundström, 2008). In 1997 these authors found that the DSDS correlated negatively with the self-reported number of responsible collisions (but not those where the informants were not responsible[22]), while the 1998 study had no effects for accidents. Unfortunately, the 1997 study did not test whether the association between accidents and other variables changed when social desirability was held constant, while the lack of effect in 1998 may have been due to the use of 'All accidents' as criterion instead of culpable ones only, which had stronger correlations with social desirability in the first study.

Concerning question 2 in the previous section, it can be noted that Jamison and McGlothlin (1973) and Donovan, Queisser, Salzberg and Umlauf (1985) did not find any differences for their social desirability scales between groups contrasted by accidents from state records, a non-difference that would be expected if the social desirability effect was on the report, and not the actual behaviour of the drivers. The same type of result was reported by Williams, Henderson and Mills (1974) for traffic offenders, using the lie scale in Eysenck's Personality Inventory. Similarly, in Smith (1976) it can be found that the scores on a lie scale were not significantly different when drivers were grouped according to state records of accidents/no accidents, but that the existing difference almost doubled when self-reported accidents were added, becoming significant at $p<.05$ (t-test). Those with no accidents had higher scores. Finally, af Wåhlberg, Dorn and Kline (submitted) reported negative associations between the DSDS and self-reported accidents, but positive ones for recorded crashes, in several samples.

Social Desirability in Driver Questionnaires

It can then be asked (question 3) whether there is any evidence regarding social desirability responses in traffic/driving inventories. As with so many other methodological problems in traffic safety research, little research is available, and most of the quaint few share a peculiar shortcoming.

Dula and Ballard (2003) reported rather strong correlations (.26–.40) between their driving anger scales and the Interpersonal Behaviour Survey Denial Scale, and similar results were found by Willemsen, Dula, Declercq and Verhaeghe (2008) with another lie scale, while Lajunen and Summala (1995) did not so find, using the EPQ lie scale versus various driving-related scales. Groeger and Grande (1996) reported correlations of .252 and .173 between two of their self-skill-assessment scales and the EPI lie scale, but this did not influence associations with

---

22   According to their own view; see Chapter 4.

other variables when controlled for. Fernandes, Job and Hatfield (2007) preferred to exclude the subjects who were high on the Marlowe-Crowne Social Desirability scale, but unfortunately did not report whether it correlated with any other measure taken. It is uncertain whether this method actually countered the suspected socially desirable responding problem, as this would seem to be a continuous variable, meaning that it is not the few who are high on this trait that create the problem, but the whole range of responses. Excluding the high responders would therefore decrease the effect, but probably not eradicate it.

For the Manchester Driver Behaviour Questionnaire, it has been reported to be only slightly influenced by social desirability, as measured by the DSDS (Lajunen and Summala, 2003), and not at all when measured by a nameless scale adopted from Paulhus (1991) by Wickens, Toplak and Wiesenthal (2008). However, the first was a difference between groups/situations study, not individual differences, which make the results inapplicable to the latter. Unfortunately, this difference has not been understood, and several researchers have cited Lajunen and Summala as evidence that social desirability is not a problem with the DBQ (for example, Darby, Murray and Raeside, 2009). Wickens et al. used psychology students as subjects. As a lie scale depends upon deception, it is not recommendable to use it on subjects that may be knowledgeable about its function.

These researchers seem to have included various lie scales because they suspected that people might not be telling the truth. However, for individual differences, this is only important if spurious associations are found between predictors and predicted variables. Strangely enough, this is also were all these studies, except Groeger and Grande (who did not test it against crashes), stop short. They have not tested if associations are attenuated. We are therefore left with virtually no evidence concerning the effects of socially desirable responding on the prediction of traffic accidents in questionnaires, apart from its effect on a few of the predictors.

Only two studies seem to have held constant the effect of social desirability when the association between crashes and predictors was computed. Conner and Lai (2005) reported that there were no effects for the DBQ and the Driver Attitude Questionnaire. However, no values were given.

The other study tested this for the DBQ, the Driver Behaviour Inventory,[23] the Driving Anger Scale,[24] a drug scale, and the Sensation Seeking Scale (af Wåhlberg, submitted). The Driver Impression Management scale of the DSDS was found to explain more than half of the shared variance between self-reported accidents and these driver inventories. It should be noted that the DSDS scale had been validated by testing it against different collision sources (af Wåhlberg, Dorn

---

23   Aggression scale; Gulian, Glendon, Matthews, Davies and Debney, 1988; Matthews, Dorn and Glendon, 1991; Matthews, Desmond, Joyner, Carcary and Gilliland, 1997; Matthews, Tsuda, Xin and Ozeki, 1999.

24   Deffenbacher, Oetting and Lynch, 1994; Knee, Neighbors and Vietor, 2001; Deffenbacher, White and Lynch, 2004.

and Kline, submitted). As expected, self-reported collisions correlated negatively with the scale, while recorded ones had a positive tendency.

These two studies therefore had starkly different results, something that cannot be explained by different populations or small samples. What differed was the lie scale; Conner and Lai used the Crowne-Marlowe instrument, which does not seem to have been validated for traffic use. However, it can also be noted that the correlations between inventories and crashes in the Conner and Lai report were much smaller than those in af Wåhlberg (submitted). It could therefore be the case that there was indeed no effect of social desirability in the Conner and Lai study (for reasons unknown). However, the very small effects that did result could still be explained by other CMV mechanisms.

Social Desirability and ADHD

When discussing various types of social desirability effects, especially on collision reporting, it can also be noted that there exists, in some situations, the possibility of finding spurious effects if some individuals are not affected by it at all, and therefore report their accidents truthfully. Such individuals could possibly be drivers diagnosed with Attention-Deficit/Hyperactivity Disorder (ADHD). In several studies, such drivers have reported vastly higher numbers of crashes than compared to controls. However, if these studies are compared to the ones using official records, a difference is seen; in self-reports, the mean difference between groups is 98 percent,[25] while for records it is 83 percent.[26] This is a fair difference in effect, and this is even more pronounced if the (self-reported) result of Weiss, Hechtman, Perlman, Hopkins and Wener (1979) is added. In that study, the difference was an incredible 1,757 per cent, a value that cannot really be trusted. For one thing, the p-value reported was <.05, but such a vast difference would yield a much lower value. It can therefore be suspected that this difference is simply an error of transcription, as the number of decimals for the means differed. The incredibility of this result does not seem to have registered with the researchers in this area, though, as it is regularly cited without comment (for example, Barkley and Cox, 2007).

Two studies (Barkley, Fischer, Edelbrock and Smallish, 1990; Barkley, Guevremont, Anastopoulos, DuPaul and Shelton, 1993) have used crashes reported by the subjects' parents as outcome, and although the differences for those are

---

25   This value was based on percentage differences of means or number of drivers with accidents in the following studies: Barkley, Murphy and Kwasnik, 1996; Murphy and Barkley, 1996; Nada-Raja, Langley, McGee, Williams, Begg and Reeder, 1997 (two values); Barkley, Murphy, DuPaul and Bush, 2002; Knouse, Bagwell, Barkley and Murphy, 2005 (two values); Barkley, 2006; Fried et al., 2006; Fischer, Barkley, Smallish and Fletcher, 2007.

26   Lambert, 1995; Barkley, Murphy and Kwasnik, 1996; Barkley, Murphy, DuPaul and Bush, 2002; Barkley, 2006.

even larger (mean 157 per cent), it is difficult to interpret these results. Could parents of children with an ADHD diagnosis share some of their characteristics (due to genetic and/or cultural inheritance), like the suspected low susceptibility to social desirability?

It is also instructive to study some other relationships between self-reported and recorded collision data in ADHD studies. In Barkley, Murphy and Kwasnik, (1996) the correspondence between self-reports and records (in terms of means) was 50 per cent higher for the ADHD group as compared to controls. This fact would also be in agreement with the hypothesis about ADHD youngsters not being influenced by social desirability. However, in Barkley, Murphy, DuPaul and Bush (2002), both ADHD and control groups reported about three times as many crashes as could be found in the records.

Furthermore, Fried et al. (2006) found that 50 per cent of drivers with ADHD reported having been rear-ended, while only 17 per cent of controls did so, which is important because this kind of accident is usually considered to be the hitting driver's fault (for example, Evans and Gerrish, 1996; Lie, Tingvall, Krafft and Kullgren, 2006). Although this may not always be true (habitual strong braking behaviour could have such an effect, see Babarik, 1968), the difference found by Fried et al. would seem to be too large to be solely due to such a behaviour and/or differential exposure. A bias in reporting is likely to explain at least some of it. Reimer, D'Ambrosio, Coughlin, Fried and Biederman (2007) also reported an association between ADHD and being shunted.

Furthermore, the use of this type of accident would seem to be a viable method for testing the suggested bias in future studies. Finally, it could of course be argued that the ADHD drivers were claiming a high percentage of shunts to try to make out that their accidents were not their fault, but then one would need to explain why they reported the crashes in the first place. Another identifiable group of drivers that could yield a similar effect (no social desirability) is aggressive ones, where Galovski and Blanchard (2002) reported that such persons undergoing treatment often seemed not to understand that their aggressive behaviour on the road was dangerous and/or socially unacceptable.

In general, the research on ADHD and road traffic accidents can therefore be suspected to have resulted in spuriously strong effects. As will be discussed in Chapter 6, these studies (and those on aggression) also suffer from other methodological shortcomings that make them unreliable.

Consistency Motif

Another possible common method variance mechanism that could explain how a spurious association between self-reported accidents and other variables could come into existence is congruent reporting (af Wåhlberg, 2002a; af Wåhlberg, Dorn and Kline, forthcoming). This hypothesis states that people have a perception about what constitutes safe driving, and reply congruently to questions about their behaviour and their accidents, as not to appear inconsistent. Associations will therefore be found between the reported accidents and the variables the drivers think are important for safety, no matter what the actual correlation is. This hypothesis is akin to the 'consistency motif' suggested for cognitions and attitudes (Osgood and Tannenbaum, 1955), in which respondents search for similarities between items and respond more consistently than they actually act in real life. The mechanism is also similar to what has been called self-generated validity (Feldman and Lynch, 1988). When concepts are presented concurrently, like in a questionnaire, to a respondent, this person might then make a connection between them that did not exist beforehand, and change the answers accordingly.

Unfortunately, no way of measuring congruence/consistency motif has been found, and its effects can therefore not be partialled out from driver inventories. However, an indirect method of investigating this mechanism has been constructed, and testing is under way. This method is described in the next section.

Testing for CMV Effects

The proposed hypotheses about common method variance effects can in principle be tested in at least five ways. First, the mechanism as such could be measured in some way differently from the variables on which it is thought to have an effect, and this effect held constant. This method is in principle applicable to social desirability, but currently there does not seem to be a way of assessing this construct that is really independent of the scales where the effect is to be held constant (because there are similarities between the items).

It can here be noted that the DSDS scale, although a definitive methodological improvement in traffic safety research, has a distinct shortcoming in that it cannot capture the socially desirable responding of groups that do not have traffic safety as an ideal. Here, there is a dire need of development, as we do not really know whether effects such as over-reporting due to social desirability can influence the results of driver inventories. Different lie scales and other methods to detect fake responding are needed. One innovative method that deserves mentioning was that of Cassel, Ploch, Becker, Dugnus, Peter and von Wichert (1996), who included items about sense of smell and vision in a questionnaire about accidents and fatigue, to see whether there was a change for these after treatment for sleep-disordered breathing. Although it is uncertain whether such a simple lie scale does work, the principle in itself is interesting.

Second, for congruent reporting, it is predicted that the predictive power of various traffic safety items versus self-reported crashes will be stronger in groups that think that the item under analysis is important for traffic safety, as compared to those who do not think it is important. In general, the hypothesis here is that it is the drivers who are high on the importance variable who carry the effect, because those who do not think so have no need to reply in a similar fashion between the behaviour and crash items. Furthermore, this means that those variables that the respondents think are important for traffic safety will be the ones that tend to correlate with accidents.

Third, the associations of predictors with different dependent variables (that is self-report and recorded data) can be compared. If common method variance is at play, the self-report data should have stronger associations with most variables (mainly those that people believe are important for traffic safety), and the pattern of associations across several predictors should differ. This method is applicable to all spurious effects, regardless of the mechanism behind them, although the difference in means between sources makes the evaluation difficult. As can be seen in Table 2.5, what little data for crashes that has been found indicate rather low agreement between effects for self-reported and recorded data. For other variables, Boyce and Geller (2002) reported that they did not find the usual associations between various personality constructs (for example, venturesomeness) and risky driving, when using objective measurements of the latter. Similarly, Verwey and Zaidel (2000) reported a correlation of .40 between extraversion and self-reported simulator driving errors, but when the self-reports were replaced by actual data from the simulator, the association shrank below .20.

Fourth, it is actually possible to test for CMV effects without measuring any of these constructs, or using crash data from any other source but the respondents. What is undertaken is to distribute the questionnaire twice (or more). First, the reliability of the replies between waves can be tested, something which is rather unusual in traffic safety research. Second, the correlations between scales and self-reported accidents can be compared. If there are CMV effects, the highest such correlation should be found within each wave of the questionnaire, as compared to between waves. Third, the scales of the questionnaires can be added and tested against the accident variables. If there are CMV effects, the resulting correlations should be lower than those within each questionnaire. If there is a true effect, the correlations should instead increase, because the mean of two (or more) measurements should be a better estimate of the true value, as random errors tend to cancel each other out between waves (Rushton, Brainerd and Pressley, 1983). This method seems never to have been used in traffic safety research.

The fifth method for testing for CMV effects is the multitrait-multimethod matrix of Campbell and Fiske (1959). Here, two methods (say questionnaires and interviews) are used to measure two different constructs, for example, political attitudes and happiness of marriage. If the correlation between the two constructs is higher within each method than between each construct as measured by each method, then there is evidence of CMV. However, this method would seem to suffer somewhat from the problem of defining how different constructs should be as to make the test valid.

As there are many CMV effects, it is very difficult to test and control for them all, especially as the available methods for their measurement are open to criticism. This problem is easily circumvented by using a different source of data for the dependent variable (that is really the third testing method suggested). If an inventory cannot predict recorded crashes, it does not have any predictive validity regarding individual differences in traffic safety. That this method makes it necessary to use larger samples, or high-risk populations, and/or long time periods, is just something that has to be accepted, although self-report researchers are likely to abhor this extra burden of work.

Concluding Remarks

It can also be noted that common method variance is affecting other areas of research too, where the same basic problem is present. Thus, Armitage and Conner (2001) reported, as the result of a meta-analysis, that the constructs of the Theory of Planned Behaviour (TPB) explained 50 per cent more variance when the outcome was self-reported, as compared to when objectively measured. Such results were also reported specifically for the TBP versus simulator speed behaviour by Elliott, Armitage and Baughan (2007). Unfortunately, such direct comparisons of explained variance can seldom be used for collisions, as the archive sources almost always suffer from lower variance than self-reports. When this is not the case, the expected difference occurs (af Wåhlberg, Dorn and Kline, forthcoming). When methods for controlling for differences in variance between accident sources are invented, these comparisons will show whether similar effects of common method variance are present in various driver inventories.

It is also apparent that in other research areas where questionnaires are used, the problem of common method variance, especially due to social desirability, is taken much more seriously than in traffic safety. Management, consumer and personality psychology, for example, are fields of investigation where the discussion about this issue is continuing.

Despite the many arguments and results that seem to indicate that something is seriously wrong with the validity and results of self-reported traffic behaviour data, it is still the most popular road user research method today. Why is this so? Is it just because it is cheap and deceptively simple to use? Because it allows the researcher to generate any kind of hypothesis and 'test' it? Or because it is so easy to follow the cookbook recipe for how to construct an inventory? What about the ease of publishing the results? Although many of these reasons may be applicable, there still seems to exist far-reaching convictions about the validity of self-reports. It can therefore be wondered where such beliefs originate, because they certainly cannot be based upon the evidence presented here.

Arguments in Favour of the Use of Self-reports

> ... across a number of social behaviours, research studies have demonstrated that significant and reasonably strong associations can be found between self-reported and more objectively measured behaviour ... we argue that there is little

reason to assume that self-reported behaviour will not serve as a good proxy for more objectively measured behaviour. Elliot and Baughan, 2004, p. 389.

Within social science, to which the study of driver behaviour must be said to belong, it is common practice to discuss the validity of the measures taken, as this is often problematic. Given the evidence presented in previous sections about the many shortcomings of self-reported driver behaviour data, it can be asked, how do researchers who use self-reports handle the awkward situation of presenting data of an inferior quality, as compared to the actual behaviour? The present section will show some examples of the most common methods used by researchers to cover up this shortcoming.

The simplest method is of course not to mention the war, that is, this part of the method is not discussed.[27] Yet another is to confess to the possible deficits and plead for mercy from the readers, which seems less common (but see point 8 below). Thereafter, several different strategies emerge when canvassing the literature:

1. The citing of a few papers that actually do support the use of self-reports.
2. The citing of reports of very doubtful relevance for the subject.
3. Coupling 1) and 2) with very positive interpretations of the results.
4. Careful avoidance of most of the papers in Tables 2.2–2.4, which show less positive pictures than the papers actually quoted.
5. Logically faulty analyses and arguments in support of self-reports.
6. Strong generalization from one finding to another, for example, speed results are claimed to be relevant for the validity of other self-reported driver behaviours.
7. Talking about the validity of measurements in general terms (mostly content, construct, convergent or discriminant-related[28] validity, but seldom criterion-related).
8. Placing the responsibility of validation upon future studies, or rather, other researchers.

1-2-3) Reference Abuse

The technique of biased referencing can be illustrated with a paper by Bener, Özkan and Lajunen (2008), who, under the heading of 'Methodological limitations' claimed that '... several studies have indicated that self-reports of driving correspond well to actual driving behaviour. For instance, Ingham (1991) found high

---

27   As practiced by, for example, Bener, Murdoch, Achan, Karama and Sztriha, 1996; Deffenbacher, Lynch, Oetting and Yingling, 2001; Deffenbacher, Lynch, Oetting and Swaim, 2002; Koushki and Bustan, 2006.

28   The terminology is that of Anastasia (1988).

correlations ...' (p. 1416). In this instance, just a single reference was provided, and this paper was also a fairly obscure one. Furthermore, when located, it was not possible to ascertain anything about the validity of self-reports from this report, as no figures of any associations were reported, and, in general, the reporting was very hazy.[29] Furthermore, as Bener et al. used the DBQ, which, as discussed in the previous section, is very poorly validated, no appeal to unspecified 'recorded driving' can actually help in this instance.

The above quote under the section of agreement between sources of accident data from the introduction in Newnam, Griffin and Mason (2008) gave one reference for its rather strong claim about self-reported crashes correlating strongly with 'objective measures of crashes' (whatever that is), and is another good example not only of biased referencing, but also the use of doubtful sources. The study thus referred twice to Lajunen and Summala (2003) for the claim that self-reported and objective measures of crashes correlate strongly. However, crashes were not included as a variable in that study. It could of course be the case that Newnam et al. used Lajunen and Summala as a secondary reference, because the subject of validity of self-reported accidents was indeed mentioned in that paper, but only in terms of the forgetting rate of drivers. So, the origin of the claimed strong correlation between self-reported and objective measures of crashes is therefore impossible to know.

Furthermore, this confusion was not lessened when similar claims were made in the discussion of the same study; '... self-report measures of crashes have been found to be strongly correlated with independent observations (Lusk, Ronis and Baer, 1995), and accurate (sic!) recall of workplace accidents has also been found to be acceptable in older age groups (25–54) for a period up to 12 months (Landen and Hendricks, 1995).' (Newnam, Griffin and Mason, 2008, p. 642). Why the Lusk reference was given is totally inexplicable, as it concerned the use of hearing protection aids. It could also be wondered how crashes could be recorded by independent observation. The Landen and Hendricks reference was correct in the sense that these authors did study the recall of at-work injuries. However, two other features of this study make it improper as a support for the use of self-reported accidents in the Newnam et al. study. First, the 18–24 year age group strongly under-reported their injuries, and although Newnam et al. did not report how many of their subjects were in this category, this is a major limitation. Second, the Landen and Hendricks study was about group rates, that is the responses of the subjects were not compared with their own records, but with national averages.

Newnam et al. also used the strong generalization technique; '... it has been shown that self-report driving questionnaires are associated with minimal social desirability bias (Lajunen and Summala, 2003).' (p. 642). However, Lajunen and Summala only tested this for the DBQ, not for any of the scales used by Newnam et al. Also, the Lajunen and Summala study was an experimental manipulation, which yielded no differences between situations. This does not mean that there

---

29   It would seem that the report is a transcript of a speech.

is not an effect within a situation, that is, between individuals. This logical error of applying a situation difference to an individual differences study has also been made by McEvoy, Stevenson and Woodward (2006), referring to the same study as Newnam et al.

Similarly faulty and biased referencing concerning the validity of self-reports can be found in, for example, Fernandes, Job and Hatfield (2007), who claimed '...relevant literature suggests that participants' self-reports in this area are reasonably accurate (for example, Aberg, Larsen, Glad and Beilinsson, 1997; Prabhakar et al., 1996; Ulleberg and Rundmo, 2002).' (p. 68). The first reference did report a study of the validity of self-reported speed, with two correlations of .36 (as described previously in the speed section). The other two, however, were not validity studies, and used only self-reported data. Why they were considered relevant is incomprehensible. Here, it can also be noted that ten different 'risky driving behaviours' were used as outcome variables by Fernandes et al., and not a single validity figure was given, the closest equivalent being 'a significant correlation' between observed and self-reported speed, found by West, French, Kemp and Elander (1993). As described previously, these correlations were probably the highest ever reported, and still accounted for only about a third of the variance. This type of biased and/or erroneous referencing is unfortunately rather common.[30]

Returning to claims of validity by agreement by sources,[31] it can be pointed out that although agreement, even if it existed to any acceptable degree, is a necessary, but not sufficient condition for acceptance of self-reports as scientifically useful. It needs also be established whether there are any systematic biases that influence the associations between variables.

2) Second-Hand Knowledge

As would seem to be the case with Newnam et al., researchers will sometimes make claims about self-reports that are based upon second-hand references, while implying that they are first-hand. For example, Malta, Blanchard and Freidenberg (2005) said that '... Panek et al. (1978) found that self-reported MVA (Motor Vehicle Accident) information was as reliable as information from police reports, insurance companies, and state agencies.' (p. 1480). However, Panek, Wagner, Barrett and Alexander (1978) did not investigate this themselves, but referred to another study: 'A self-report questionnaire was used because as indicated by

---

30   See for example, Tranter and Warn (2008), who referred to Wallen Warner and Åberg (2006) for links between attitudes and accidents.

31   For example, 'The accuracy particularly of self-reported crash involvement has been the subject of several studies ... with the findings generally supporting a reasonable level of agreement between self-reported and official crash data.' (Langford, Methorst and Hakamies-Blomqvist, 2006, p. 577).

McGuire (1973), self-reports of accidents and moving violations are often more reliable and valid than police, state, or insurance data.' (p. 355).

Turning to the original source, McGuire did compare state records and self-reports, and regarded the first as highly inaccurate and biased. However, it appears he started out assuming that self-reports were true, and interpreted any discrepancy between these and records as evidence of errors in the latter. Neither was there any information given about how the information was matched between sources, or if there were accidents in the records that the subjects had not reported. As should be evident from the previous sections, the assumption of accuracy of self-reports is clearly faulty, while the lack of information about matching could be due to this being undertaken on a numbers and not details basis. This means that erroneous self-reporting was not detected, and the possible under-recording by some participants would have decreased the strength of the association between the variables substantially. Unfortunately, no correlation between sources was reported.

Turning to the claim about insurance sources made by both Malta et al. and Panek et al., this again is taken from another source (because McGuire did not investigate this himself, but reviewed other studies); Burg (1967). Here, it can (finally!) be found that Burg compared state and insurance records (but not self-reports), and the reference is therefore irrelevant for the subject of validity of self-reported accidents. That one source has low validity does not prove that another one is better.

In this example, it can be seen not only how a second-hand reference is referred to as first-hand, but also how the claim has been strengthened between sources, because the important word 'often' used by Planek et al. was dropped by Malta et al., who otherwise seems to have copied the original claim verbatim. In this way, erroneous information can be spread, because few people check the references given in an article. The reviewers of the Malta et al. paper obviously did not. By following this chain of distortions of facts from the original sources, it can be seen how a sort of urban myth about self-reports can start to grow. The end product will be a general acceptance of self-reports as valid, which is mainly due to ignorance.

3) Accepting Low Quality

The old saying about different people seeing a glass as half full or half empty would seem to be applicable to the self-report method. Given the research reviewed in the present text, and the arguments proposed, it is rather strange that the majority of traffic researchers still not only use self-reported data, but actually claim that they have enough validity to serve in scientific research. For example, Roberts, Vingilis, Wilk and Seeley (2008) thought that '... the level of concordance between self-reported MVCs and injuries was high ...' (p. 561) for the results of Begg, Langley and Williams (1999), where only 49 per cent of self-reported accidents could be matched to recorded ones. Also, Haglund and Åberg (2000) reviewed existing results on validation of speed reports and found them to be .27 to .65. From these correlations, they came to the view that 'Thus, the results imply that self-reports can be used in the modelling of driver speed choice ...' (p. 178). Similarly, West, French, Kemp and Elander (1993) concluded that 'The results indicated that self-reports of speed could be used as a surrogate for direct observations of speed.' (p. 554) and this claim was probably in its turn cited[32] by Lajunen and Parker (2001), presumably as an argument for their method. Furthermore, Parker, Manstead, Stradling and Reason (1992) claimed (talking about a correlation between accidents and violations) that 'Because there are no grounds for thinking that individuals would inflate (or avoid reducing) self-reported frequency of accidents, this positive association enhances confidence in the accuracy of self-reports of violations.' (p. 121). But, as we have seen, there is both reason to think, and data supporting that, inflating is exactly what some people do, regarding both these variables. The assumption that under-reporting is the only problem of self-reports (for example, Panek, Wagner, Barrett and Alexander, 1978; Laapotti, Keskinen, Hatakka and Katila, 2001) is obviously false.

5) Logical Errors

Somewhat embarrassingly, many researchers in traffic safety seem to be unaware of fairly simple facts of how various associations in data can be interconnected, and what conclusions can be drawn from one fact regarding another feature. Actually, they often seem not to understand that there might indeed be a difference between concepts.

Such faulty logic can again be exemplified by the Malta et al. study; 'It is also unlikely that the same individuals who described fairly severe aggressive episodes (biting, choking, assault) in a face-to-face interview would under-report their aggressive driving on an anonymous questionnaire.' (p. 1480). This may well be true, but it does not constitute evidence that either report is true. Neither does it tell us anything about those who did not want to admit to aggressive acts.

Another faulty argument can be found in Sullman, Meadows and Pajo (2002) in favour of their use of the Driver Behaviour Questionnaire (DBQ): 'Walton (1999a) compared the self-reported speeds of truck drivers with the mean speeds observed by the Land Transport Safety Authority (LTSA). Walton (1999a) found that the truck drivers accurately reported the speeds at which they travelled on the road.' (p. 230). However 'accurately' only pertain to the mean in Walton's data, as this was the only possible comparison. But the DBQ uses association

---

32   It is uncertain exactly what Lajunen and Parker were citing, because they did not give a page reference, and the exact quote 'Authors concluded that 'self-reports of certain aspects of driver behaviour can be used as surrogates for observational measures' (West et al, 1993).' (Lajunen and Parker, 2001, p. 254), has not been found in West et al.

measures on the individual level, and as a correlation between sources might be very low despite the group means of the variables being exactly the same (and vice versa), the Walton study results are not applicable for the instance in whose favour they were cited. The same logical error seem to have been made by Gulliver and Begg (2007), who argued that there was a high level of agreement between the self-reports of their cohort and police records. How this agreement had been calculated was not stated, but it can be suspected that it was the means that were compared.

Overall, the Sullman et al. study features several of the self-report-researcher biases described; the authors cited four references in favour of the validity of their undertaking (all of them discussed at various places in this text), and none of them actually carries any weight at all concerning this question. Furthermore, Sullman et al. did not refer to almost all the papers cited in the sections on accidents and speed in the present work, most of which were available at the time of their writing their own article.

In the vein of using mean levels as evidence of individual differences validity, Newnam, Griffin and Mason (2008) compared the percentages of drivers in the population who reported being accident involved with the data of a leasing agency who supplied their vehicles. By reaching similar percentages in both sources, they claimed 'some support' for their self-report method.

With some possible similarity, Jelialian, Alday, Spirito, Rasile and Nobile (2000) claimed that 'The consistency across samples lends credibility to the reliability of self-report data on motor vehicle-related injuries.' (p. 90). The only explanation to such a statement would seem to be that the authors were thinking in terms of mean levels, not individual differences. But as the analysis was about the latter concept, the argument (or whatever it was) was simply not relevant.

Sometimes, the faulty logic takes on the form of no argument at all, but only a statement, as shown in Iversen and Rundmo (2004): 'This study indicates that self-reports can provide a sensitive and appropriate instrument for the study of the association between attitudes, risk behaviour and accident involvement.' (p. 570). However, the study was not about this at all, but the usual approach, relating various self-reports to other self-reports. Why this would constitute evidence concerning issues like sensitivity and appropriateness was not explained.

8) 'In future research ...'

The technique of placing the responsibility elsewhere seems to indicate that even some researchers using self-reported data, mainly accidents, feel a certain awkwardness about this method. The ingenious solution is to suggest that 'future studies' should use objective data.[33] However, this intention seldom seems to be put into practice, at least not by the researchers suggesting it.

Concluding Remarks

It might of course be argued that the examples given in this section are too few to constitute evidence that there is a general problem of this type within traffic safety research. Unfortunately, the examples presented are but a few, randomly selected, instances of the very many found. It can therefore be concluded that the arguments in favour of self-reports for individual differences research in traffic safety are severely deficient.

Discussion and Conclusions

> A more promising avenue of accident prevention would be the direct, objective study of driver behaviour rather than the use of questionnaires which are indirect and probably invalid measures out of context with the driving situation. Williams, Henderson and Mills, 1974, p. 107.

It appears that, with one or two exceptions, all studies reviewed here which have tested the validity of self-reported traffic data in some way have yielded fairly low figures, and self-reports of such variables, at the individual level, must therefore be deemed to be very unreliable. Despite this, and the problem of CMV, the confidence in questionnaires abound, and any number of studies can be referenced which contain nothing but self-reported data,[34] a phenomenon that is not new (see for example, Suchman, 1970), but apparently on the rise.

Why do so many researchers use self-report data only, when the risk of common method variance is so high? It must be concluded that, apart from the disinterest in the accident variable, the main reason is that it is easy to gather self-report dependent data. Self-report research is really cheap in all senses of the word.

What kind of effects will the various biases reported above have on practical research? First, the number of studies where spurious results can be suspected is staggering. The papers discussed in the present chapter are only a part of the total, as mainly journal papers have been studied. Reports, book chapters and conference publications have largely been excluded, because they are not as accessible as journal papers. However, there is no reason to believe that these categories of

---

33   For example, Galovski and Blanchard, 2002; Ulleberg and Rundmo, 2002; Lajunen and Summala, 2003; Fernandes, Job and Hatfield, 2007; Moore and Dahlen, 2008.

34   For example, Parker, Manstead and Stradling, 1995; Parker, Stradling and Manstead, 1996; Lajunen, Corry, Summala and Hartley, 1998; Rimmö and Åberg, 1999; Taubman-Ben-Ari, Mikulincer and Gillath, 2004; Maxwell, Grant and Lipkin, 2005.

studies would be of any better quality. In general, we therefore need to start all over again in several areas, notably regarding driver aggression, where not a single study using objective data has been found.

What methods of research should be avoided in accident prediction studies, given the problems discussed here? As noted in the section about common method variance, there is a very simple solution to this problem; the use of outcome variables from other sources. When the aim of the data-gathering is slightly different, for example, holding exposure constant, it is also a good solution to use another, preferably objective, source of data instead of self-reports. Overall, the recommendation is therefore to abstain from using self-reports, as they are simply not reliable, if there is any alternative available.

It is interesting to note that hardly any research has been found which has tested the validity of insurance company data against other sources, despite being very important. Triangulation of self-reports, state records and insurance claims would be a strong methodology when studying the crashes of private drivers. In the near future, the use of black box technology in vehicles may also resolve this situation, as it will be possible to measure speed changes and other variables, where very sudden and strong changes will denote impacts.

There is actually another kind of self-report that is seldom used (for reasons unknown); concurrent reports, that is driving diaries. This method is of course limited by the perceptions, truthfulness and conscientiousness of the drivers, as well as their information being limited to their immediate environment. However, the only practical limitation seems to be that it is not as cheap, fast and easy to administer as the questionnaire. Some examples of the use of driver diaries can be found in Underwood, Chapman, Wright and Crundall (1999).

It could of course be argued that the weak correlations between self-reported and recorded accident data are to a large degree due to the imperfections of the most common criterion; state records, which are well known to have any number of biases and short-comings.[35] However, as shown, the extremely low correlations between these sources simply cannot be due to only the records being biased towards severe accidents. Therefore, the most positive conclusion to be reached is that more research is needed on this problem, especially concerning over-reporting, an effect which most researchers seem to have discounted. Also, the various common method variance mechanisms in driver behaviour inventories need to be investigated, especially in relation to crashes.

A general principle of all self-reports seems to exist; the closer in time that the report is to the behaviour, the better correspondence, probably due to memory factors (as found for accidents by Chapman and Underwood, 2000). Therefore, reporting periods should be very short (a few months at most), if memories of

---

35   For example, Zylman, 1972; Shinar, Treat and McDonald, 1983; Maas and Harris, 1984; Fife and Cadigan, 1989; Harris, 1990; Hopkin, Murray, Pitcher and Galasko, 1993; Rosman and Knuiman, 1994; Austin, 1995a; 1995b; Alsop and Langley, 2001; Rosman, 2001; for a review see Hauer and Hakkert, 1988.

driving should not be severely distorted. However, this principle is not possible to apply to crashes, as they are so uncommon as to make the variance within such a time period too small for any practical use.

One thing is certain; people can forget, or at least abstain from reporting, the most remarkable things. Kunkle (1946) told the amazing story of a pilot who after an interview about his accidents returned to the researcher and said that he had just recalled that he had killed a ground crew man by hitting him with the propeller. Apparently, even the messiest incidents may be at least momentarily forgotten.

Overall, the state of the art for traffic safety research in the early 21st century would seem to be that the majority of researchers are making grandiose claims from data of low or unknown quality, disregarding the fairly large and mounting evidence of the dubious validities of their central variables. Although many pay lip service to the problems of socially desirable responding and forgetting, this is only used to bolster the claims that the associations found must be real. No one seems to have understood the significance of CMV effects (this term does not even seem to be known by traffic safety researchers). Finally, the practical uses of the many existing traffic behaviour inventories may be limited, not to mention how scarce the published evidence for any successful use of their results is, for example, in guiding driver training. But then again, if the evaluation of an intervention is made with a questionnaire, it is fairly certain that some positive effects will be found. What these effects are due to, on the other hand, is very much less certain.

Today, few quantitative studies within social science or medicine would be accepted if they did use a method where evaluators or experimenters were not blind to the status of the subjects. However, this has not always been the case, and numerous examples can be found within the safety literature, where the results have probably to a large degree been due to researcher effects or biases (for example, Adler, 1941; Shaw and Sichel, 1961; Shaw, 1965). In those early days, there was apparently little knowledge about the experimenter effects, but with time, we learned the hard way. It is also hoped that we will also be able to learn about the pitfalls of self-reports.

Meantime, it can be agreed with Åberg and Wallén Warner (2008) that 'The result in itself is interesting as it points to the fact that self reports might contain information that is not available in objective data.' (p. 29). Whether this information is truth or fiction is to some degree still to be determined, but the existing evidence would seem to indicate the latter.

Chapter 3
Accident Proneness

But, studies tend to focus on an extremely unreliable criterion measure – accidents ... Dorn, 2005, p. 431.

Introduction

Accident data lacks stability ... Iversen and Rundmo, 2002, p. 1261.

The History of Accident Proneness

Analysis soon made it clear that a considerable proportion ... of accidents were caused by a relatively small percentage ... of the men. Slocombe and Brakeman, 1930, p. 29.

One of the most basic problems we encounter about crashes and individual traffic accident records are whether they are due to stable or transitory individual factors, or rather how much these factors contribute. In other words, do individuals cause accidents due to some inherent trait (which is fairly stable over time), or as reactions to incidents in their environment? If the first is a strong factor, then the number of collisions of individual drivers should be similar in different time periods within the individual, but different between drivers, that is a stable individual differences variable.

In this section, the available evidence for stability of collision record will be reviewed, and related to the practices of individual differences research. The reason for the interest, mainly in stability over time of accident record, by traffic safety researchers, is partly theoretical, partly methodological. Of course, the practical implications are also important.

Early in accident research, the theory[1] of 'accident proneness' was suggested as an explanation of the observation that some people had more accidents than could

---

1   Actually, accident proneness has never been comprehensively stated in a scientific way (as defined in af Wåhlberg, 2001), but the scattered thoughts about it in the papers referenced do have some theoretical properties, like the prediction of stability of accident record over time.

be expected by 'chance'.[2] This concept was the notion that some people are more prone to causing all kinds of accidents, and that this trait is stable over time[3].

Therefore, three positive associations were expected to appear in the accident records of individuals:

1. A correlation for the same type of accident between different time periods within an environment.
2. A correlation between different types of accidents within a certain time period and environment.
3. A correlation for the same type of accidents between environments within a certain time period.

In summary; an 'accident prone' person was expected to have a larger share of accidents as compared to the mean of the population, regardless of situation and timeframe. However, the finding of low correlations between accident records in different time periods (for example, Forbes, 1939; Stewart and Campbell, 1972) seems to have led many researchers to assume that traffic accidents is an unstable variable, that is accident proneness is a very weak factor at best[4] (for example, Arbous and Kerrich, 1951; Harrington, 1972; Maycock, 1985; Summala, 1988). Indeed, the concept of accident proneness has been criticised, refuted and found faulty by many[5] while being defended by very few (for example, Rawson, 1944). Also, the testing of the theory by comparing actual and 'expected' distributions was a blind alley[6] that contributed nothing but disagreement. Today, hardly any research about accident proneness and/or stability of accident record is published (see Tables 3.1 and 3.2).

Other positions on the question about why there are differences in individual accident records is that having an accident will make you more cautious, thus preventing further ones (Maag, Vanasse, Dionne and Laberge-Nadeau, 1997),

---

2   Greenwood and Woods, 1919; Newbold, 1927; Slocombe and Brakeman, 1930; Farmer and Chambers, 1939; for reviews, see Froggatt and Smiley, 1964; Porter, 1988.

3   However, different researchers have interpreted this basic concept very differently (see McKenna, 1983, for a discussion).

4   This conclusion was sometimes due to a very strong interpretation of accident proneness, where a small percentage of drivers were expected to have most of the accidents all the time (for example, Slocombe and Brakeman, 1930). When this extreme interpretation of the accident proneness concept was not borne out, this was seen as evidence that the concept was wrong (for example, Cobliner and Shatin, 1962).

5   For example, Brown and Ghiselli, 1948b; Adelstein, 1952; Kirchner, 1961; Cameron, 1975; Bernacki, 1976; McKenna, 1983; Mohr and Clemmer, 1988.

6   This part of the accident proneness controversy will not be treated here, as it is mainly a problem of statistical theory, not of practical research. The interested reader is instead directed to the following references: Mintz and Blum, 1949; Maritz, 1950; Blum and Mintz, 1951; Mintz, 1954a. See also the criticism of the distribution-fitting research tradition by Sass and Crook (1981).

and that accidents happen to individuals in short periods, with safe periods in between (Cresswell and Froggatt, 1963; Blasco, Prieto and Cornejo, 2003). None of these hypotheses predict a stability of accident proneness over time, or types of accidents, or environments.

Methodologically, it is interesting to note that most researchers who have paid any heed to accident proneness seem to have criticized the concept, while hundreds of studies have been published that implicitly accept it. The studies referred to are those about prediction of individual crash record, dozens of which have already been referenced in the preceding chapters (with more to come). The majority of these have tested a fairly stable type of variable (personality, driving behaviour, socio-economic standing, education, and so on) as predictor of traffic safety. But why would such variables predict accident record if the latter was not stable, at least for the time period used for gathering of crash data (usually a few years, af Wåhlberg, 2003a)? If an individual's personality, for example, predicts high proneness, this cannot be true at one time and untrue at another. Unstable variables (divorce, the death of a relative, and so on) may predict an increased accident liability (or proneness?) for a short period of time, but not for longer periods, unless they are indicative of something else, which is stable over time.

Accident Proneness and Traffic Safety Research Methodology

If a stable trait of accident proneness could be shown to exist, the methodological consequences would in some ways not be very revolutionary for current research practice; most traffic safety researchers act as if they accept a fair amount of stability in drivers' accident records (while at the same time often blaming 'the unreliability of accidents' for their weak results). One problematic method can be pointed out, however, because the assumptions underlying its use (if there are any, they have never been discussed) would seem to be very different from accident proneness. This is the habit of using a variable time period for different drivers when constructing the accident criterion variable (this will also be discussed in Chapter 6). What is a researcher thinking when he/she asks a group of drivers about their 'number of accidents' without specifying any time period? And what is actually being measured when the age range in the group is more than forty years, meaning that different drivers will report about their crashes for vastly different time periods? As no researcher using this method seems to have discussed it anywhere, it can only be guessed that the logic is something along the lines of a certain number of accidents being a driver's allotted quota, and this number will be reached fairly quickly, where after there are very few crashes. How this 'logic' works together with a current measurement of some individual characteristic of the driver (for example, aggressiveness) is more difficult to gauge, however. If drivers caused accidents due to aggression early in their careers, then they need to still be aggressive when measured (otherwise they would not be labelled aggressive). But why then does this aggression not cause accidents continuously? The 'variable time period' problem is for these reasons one of the big logical mysteries of traffic safety research, and an explanation from those who use this method is critical to our understanding of it.

Another methodological consequence of the existence of a trait of accident proneness is the increased importance of culpability for accidents. As discussed in Chapters 1 and 4, few researchers have paid any heed to this aspect when doing accident prediction studies, and it was claimed that this is a methodological shortcoming. The accident proneness concept states that some people are more prone to *causing* mishaps, that is, being culpable for what happen to them. Interestingly, this facet of accident proneness has never been exploited, and hardly even discussed. Regarding stability over time, the prediction would be that stronger associations would exist between different parts of accident records (time and/or environments) for this subgroup of incidents. However, most of the researchers who have included separate calculations for this have come to the opposite conclusion (for example, Peck, 1993), as their coefficients were lower for the latter, as compared with those for 'All accidents'. However, they have never raised the question of whether their assignment of culpability has been correct.

The question about culpability ties in with another important feature of accident proneness that has rarely been discussed; exposure to risk, because both are about how much of an individual's accident record is due to a stable trait, and what is contributed from other sources. Although lately researchers seem to have had little interest in this, the basic concept of accident proneness is about unequal accident numbers within a similar environment, for similar time periods. The accident proneness concept is not a denial of the importance of quality and amount of exposure, but about differences in individual proneness given equal opportunities.[7]

Hypothesis

It will here be suggested that the interpretation of low correlations between accident records in different time periods as 'low reliability' in the meaning of a varying behaviour has been thoroughly wrong, and due to the application of the concept of reliability upon data for which it was not suited. In essence, a low stability coefficient in crash data can be due to fluctuations in behaviour between time periods and/or restrictions in the range of the distribution.

Increasing the length of time periods used for stability computations has been noted by several authors as having a positive influence on the coefficients found, while the mean number of accidents (or rather the variance, but this value has very seldom been reported) has mainly been commented on in the negative; given such

---

7   It could of course be argued that some people create or seek out very dangerous environments (young people who go nightclubbing instead of playing chess at home), and that it therefore becomes difficult to see what is the environment and what is the individual.

low values, low intercorrelations are bound to result. However, this line of thought does not seem to have been systematically pursued.

It has been proposed that the variance and mean of a predicted variable sets a limit to the amount of variance that is possible to predict in data (Newbold, 1926; Cobb, 1940; Peck, 1993; Gebers and Peck, 2003[8]). This axiom was developed for the relation of accidents to various predictors, but may as well be applied to associations in accident records between time periods. The meaning of this is that it is not mathematically possible to find strong correlations in the data used in most studies, as they have typically utilized car drivers and short time periods, equalling low means and variances. Also, it might be pointed out that a correlation (the usual statistical method in stability studies) does not take into account the stability inherent in all the zeroes; most car drivers do not have any accidents in one year, and none in the year after too, which is an amazing stability of accident record. For example, 80 per cent of the elderly drivers studied by Daigneault, Joly and Frigon (2002) remained accident-free for a period of six years. Therefore, the results concerning accident record stability ('accident proneness') have all been artificially deflated, due to low means and variances.

So far, conclusions about the stability of accident record over time (usually labelled as accident proneness) have been reached based upon qualitative reviews. No one has ever analysed the available data quantitatively, even to calculate a mean correlation.[9] However, due to the restricted variance of accident data, a mean would not really be a very informative result, and would rather gloss over the important systematic differences instead of highlighting them.

To bring out the suspected systematic effects in motor vehicle crash data (a strong stability over time), a different method is therefore needed, which relates the size of the correlation to the (restricted) amount of variance. The solution is a variation of the secondary correlation method described in the introduction, whereby the size of the correlations between time periods can be correlated with other features of the studies. With such a tool, a meta-analysis of available research on traffic accident record stability over time could be undertaken, and yield results that are not influenced by the problem of the restricted variance. Furthermore, it could be investigated whether the effect of culpability was misinterpreted in a similar way to the basic stability associations, and whether control for the amount of exposure can increase the strength of the associations between time periods.

Thereafter, what little research has been found about stability of accident record (over time and between categories) for other types of mishaps was reviewed, as this is also important for the evaluation of the accident proneness concept and safety research in general. Finally, the equally small amount of research available

---

8   The formula is known as the Newbold-Cobb reliability index: $R = \sqrt{(S^2 - X)/S^2}$, where s^2 = Variance of accident rate, and X = mean of accident rate. Adapted from Peck (1993).

9   Salgado (2002) reported an 'average reliability' of .45, but gave no details about what studies had been included. This was probably for workplace accidents.

regarding the influence of transitory individual factors on accident record was summarized, as this in many respects is the opposite of the stable part of individual accident record.

Meta-Analysis of the Stability of Traffic Accident Record Over Time

> In longer periods these chance factors do not play so important a part, so that the longer the period of exposure the more suited it is for measuring the stability of accident proneness. Farmer and Chambers, 1939, p. 12.

The aim of the meta-analysis of stability of driving record over time presented here was to show that the main determinant of the size of a correlation between (traffic) accidents in different time periods is the variance and/or the mean in the samples. This means that there was also an expected association with the type of population (professionals usually have higher crash rates per year) and the time period used, but only because they would influence the means and variances. However, as the variance has seldom been reported, the meta-analysis used the mean of the samples, which probably correlated highly with the variance.

Data was sought as described in the introduction, but was mainly achieved by searching reference lists in relevant studies. There is a fair amount of evidence concerning the inter-correlation of driver accident record between time periods. This research is summarized in Tables 3.1–3.2 concerning some variables, which could be thought to have some impact on the results. Starting with the source of data in each study, it should be remembered that different ways of gathering data have different validities, as discussed in Chapter 2, which could influence intercorrelations. Unfortunately, very few self-report studies have tested the stability over time in their collision data, and concerning company versus state data there is a large overlap with the type of population investigated (subjects). Regarding the latter, it could be suspected that professional drivers' exposure is much more homogenous concerning both quality and amount, and that the intercorrelations therefore would be stronger than those of private drivers. Also, there were two professional groups with large enough numbers for separate analysis; bus and tram drivers. The number of subjects in a study influences the accuracy of the result as a population measure, and it may be noted that most car driver studies have used very large Ns, several hundred times larger than professional driver ones.

The time period and mean are of course intimately intertwined, and could be suspected to be the most important determinants. However, it should be remembered that a very long time period can be expected to negatively influence the stability of accident-causing behaviour. The stability between time periods in samples with equal means should therefore be stronger for those with shorter time periods.

Meta-Analysis

> In most of these studies, less than 10 per cent of the variance in crash rates was accounted for. This implies that transient factors must play by far the most important role in crash causation. Elander, West and French, 1993, p. 281.

A total of 78 different samples (displayed in Table 3.1) gave values for time periods, accident correlations between these and mean number of accidents in the sample. First, it can be noted that the mean number of accidents per year differed between car drivers (0.063, N=30), bus drivers (1.138, N=41) and tram drivers (1.321, N=7). Correlations were squared to create a linear function, that is, they were replaced by values for amount of variance explained. Thereafter, the associations between the variables described were calculated, as shown in the second row of Table 3.3.

It can be seen that the mean number of accidents correlated strongly with the squared time intracorrelations (range of r .03 to .77), as expected, while number of years used (first and second period, and total) for calculations were much weaker predictors within the total sample (total period 0.5 to 12 years). Furthermore, an outlier problem was detected (see Figure 3.1). One value (.77 from Blasco, Prieto and Cornejo, 2003) was extremely different in terms of number of accidents versus correlation, and was therefore deleted from further analysis.[10] Without this outlier, the association for mean number of accidents increased strongly, as can be seen in the third row of Table 3.3.

Thereafter, separate sets of correlations were calculated for each separate driver population where there was enough data for this; car-, bus- and tram-drivers. Again, an outlier problem was detected, this time for car drivers. The correlation from Burg (1970) and the value for males in Harrington (1972) did not match the overall pattern, and strong increases in correlation resulted from omission of these values (data not shown).[11]

It could be argued that part of the reason for the extremely strong correlation between accident association and mean number of accidents was due to many of the data points being interdependent. For example, the data from Häkkinen (1958) overlapped totally for time periods within two of the groupings, which in their turn were part of the larger sample. However, taking one value only from each of the

---

10   No apparent explanation for this outlier could be found, although there seems to be two possibilities; a change in bus company accident data procedures during the measurement period, which is not very likely, as the other value from this study fitted with the other, or a strong curvilinearity of the association. As there is a lack of published data on long time periods/high risk populations, whether the second explanation is correct cannot currently be ascertained.

11   The reason for the discrepancy in the Harrington values could be a simple error in transcription, as both values from this study were the same.

Table 3.1 The results from studies on the association between numbers of traffic accidents in different time periods, which were included in the meta-analysis

Study	Source of data	Subjects	N	Time periods (years)	Correlations, all accidents	Mean, all accidents, total time period	Correlations, culpable accidents	Mean, culpable accidents, total time period	Comments
Bach, Bickel and Biehl, 1975	Probably company records	Bus- and tram drivers	35	4/4	.76	11.44	–	–	–
Blasco, Prieto and Cornejo, 2003	Bus company records	Bus drivers	2,319	1/1 4/4	.55 .77	8.4 33.6	–	–	One-year period calculated as the mean of all correlations between the eight single years
Burg, 1970	State records	Car drivers	7,841	3/3	.134	0.254	–	–	The same data can be found in Burg, 1974
Creswell and Froggatt, 1963	Company records	Bus drivers	183 244 708 106 114 79	2/2 2/2 2/2 2/2 2/2 2/2	.248 .297 .236 .231 .244 .420	3.9945 4.4467 2.2924 2.2075 2.7281 3.5823	–	–	Results calculated from tables 6.16–6.21
Daigneault, Joly and Frigon, 2002	Insurance company records	Mainly car drivers?	187,620 131,334 71,637 35,818	3/3 3/3 3/3 3/3	.0931 .09745 .10106 .11240	0.244 0.259 0.300 0.346	–	–	Elderly drivers
Farmer and Chambers, 1939	Company records	Omnibus and trolley bus drivers	166 (A) 398 (B) 86 (C) 67 (D)	1/1 1/1 1/1 1/1	.251 .182 .174 .194	3.16 3.9 1.85 2.8	–	–	A, C and D groups values calculated as the means of intercorrelations between single years in table III. Means of accidents from table II
Forbes, 1939	State records?	Mainly car drivers?	29,531	3/3	.106	0.239	–	–	Correlation calculated from the raw data presented in the original paper

Table 3.1 *Continued*

Study	Source	Driver type	N	Period	Value 1	Value 2	Value 3	Value 4	Notes
Gebers, 2003	State records	Mainly car drivers	190,805 184,836 169,794 162,582 118,025	3/2 3/3 3/5 6/3 6/6	.062 .074 .096 .100 .131	>0.2556 >0.3034 >0.3966 >0.4478 >0.6104	(.035) (.051) (.055) (.063) (.074)	>0.1439 >0.1920 >0.2867 >0.2252 >0.3656	Not possible to calculate exact means from data in the report. Culpable accidents in first period were used to predict all accidents in second
Gebers and Peck, 1994	State records	Mainly car drivers	160,525 152,931 139,485 133,745 114,618	3/2 3/3 3/5 6/3 6/6	.073 .091 .101 .106 .130	>0.2645 0.3223* >0.4171 >0.4617 >0.5707	(.041) (.051) (.058) (.061) (.078)	>0.1442 >0.1971 >0.2964 >0.2277 >0.3617	*This value was given in Gebers (1998)
Harrington, 1972	State records	Car drivers	8,126 5,789	2/2 2/2	.04 .04	0.640 0.345	– 	0.114 0.052	Males, young drivers. Females, young drivers
Hauer, Persaud, Smiley and Duncan, 1991	State records	Mainly car drivers	65,997	2/2	.088	0.306	Not in this calculation	–	Correlation calculated from raw values in table 3 in report
Häkkinen, 1958	Company records	Bus drivers	322 141 101	1/1 2/2 3/3	.361 .454 .355	3.19 5.31 6.80	None	–	Accidents due to technical circumstances excluded
Häkkinen, 1958	Company records	Tram drivers	363 363	1/1 2/2	.233 .470	2.04 4.16	–	–	–
Häkkinen, 1958	Company records	Bus drivers	52 52 52 52	1/1 2/2 3/3 4/4	.357 .534 .507 .577	2.36 4.42 6.72 7.99	.284 .145 .385 .433	0.75 1.48 2.04 2.44	Culpable accidents were about half of all for bus drivers, and 1/3 for tram drivers. Partial overlap with samples above
Häkkinen, 1958	Company records	Tram drivers	44 44 44 44	1/1 2/2 3/3 4/4	.231 .467 .533 .674	3.45 6.18 7.92 11.18	.069 .016 .048 .299	0.80 1.43 1.59 2.44	–
McKenna, Duncan and Brown, 1986	Company records	Bus drivers	91	1/1	.30	2.65	–	–	The need to exclude non-culpable incidents was discussed. Probably Pearson correlations

Table 3.1 Continued

Study	Source of data	Subjects	N	Time periods (years)	Correlations, all accidents	Mean, all accidents, total time period	Correlations, culpable accidents	Mean, culpable accidents, total time period	Comments
Milosevic and Vucinic, 1975	Company records	Tram drivers	134	2/2	.48	4.79	–	–	Mean for four years calculated from that of five years
Moffie and Alexander, 1953	Company records	Truck drivers	100	0.25/0.25 0.5/0.5	–	–	.004 .027	0.580 0.860	Preventable accidents. No definition given
Peck, 1993	State records	Mainly car drivers	–	6/5 8/3 3/3	.216 .170 .168	–	.137 .128 .133	–	Original data in Peck and Gebers (1992)
Peck and Kuan, 1983	State records	Mainly car drivers	87,908	3/3	.100	0.336	–	–	Mean estimated from a similar, probably overlapping, sample in Table 1
Peck, McBride and Coppin, 1971	State records	Mainly car drivers	86,726 86,726 86,726 86,726	1/1 1/1 2/1 1/1	.054 .036 .060 .050	0.1809 0.1673 0.2598 0.1714	–	–	Males. Probably Pearson correlations
Peck, McBride and Coppin, 1971	State records	Mainly car drivers	61,280 61,280 61,280 61,280	1/1 1/1 2/1 1/1	.028 .041 .041 .028	0.0865 0.0824 0.1260 0.0831	–	–	Females. Probably Pearson correlations
Stewart and Campbell, 1972	State records	Car drivers	2,502 240	2/2	.0893	0.252	–	–	Values were similar when the sample was stratified for age. Spearman correlation
West, Elander and French, 1992	Self-reports	Car drivers	433 (433)	1/2 1/2	.08 (.127)	0.46 (>0.342)	– (.194)	– (>0.194)	The figures in brackets were calculated from 2×2 tables (9a and c) in the report (for the same sample as the ones above) and concern All/active accidents. As these data were dichotomised, values are probably underestimations. No explanation has been found for the discrepancy in correlations

Table 3.1 Concluded

Study	Data source	Type	No. of subjects	Time period	Correlation	Mean	Correlation	Mean	Comments
af Wåhlberg and Dorn, 2009a	Bus company records	Bus drivers	465	1/1	.279	1.90	.210	0.94	Means of ten intercorrelations between single years
			628	1/1	.239	2.04	.165	0.98	
			141	1/1	.339	2.47	.261	0.99	
			460	1/1	.208	1.78	.177	0.95	
			518	1/1	.295	1.77	.201	0.87	
af Wåhlberg and Dorn, 2009a	Bus company records	Bus drivers	465	2/2	.452	3.98	.359	2.02	–
			628	2/2	.384	4.26	.281	2.07	
			141	2/2	.536	4.93	.264	1.92	
			460	2/2	.367	3.67	.305	1.97	
			518	2/2	.423	3.58	.285	1.75	
af Wåhlberg and Dorn, 2009a	Bus company records	Bus drivers	465	2/3	.475	4.74	.420	2.35	–
			628	2/3	.429	5.11	.331	2.45	
			141	2/3	.593	6.17	.369	2.48	
			460	2/3	.390	4.45	.347	2.37	
			518	2/3	.470	4.44	.353	2.18	
af Wåhlberg and Dorn, submitted	Bus company records	Bus drivers	180	1/1	.140	0.56	–	–	The 1/1 and 2/2 periods were means of all intercorrelations
			180	2/2	.213	1.13			
			180	4/4	.267	2.26			
af Wåhlberg and Dorn, submitted	Bus company records	Bus drivers	261	1/1	.122	0.46	.108	0.40	The 1/1 period was a mean of all intercorrelations
			261	2/2	.190	0.91	.151	0.68	

Note: Shown are the data sources, type and number of subjects, time periods used for comparisons, results for all accidents and the mean of accidents in the population, and results and mean for culpable accidents only (if applicable). The time periods were mostly consecutive. Question marks denote uncertain reporting. Figures within brackets not included in the analysis. All results are Pearson correlations, with the exception of Stewart and Campbell (1972) and some uncertain ones (see comments).

Table 3.2   The results from studies on the association between numbers of traffic accidents in different time periods that did not report values that could be included in the meta-analysis

Study	Source of data	Subjects	N	Time periods (years)	Analysis	Results, all accidents	Mean, all accidents, total time period	Comments
Foley, Wallace and Eberhard, 1995	State records	Probably only car drivers	1,791	2/3	Odds ratio	2.0	0.137	Older drivers
French, West, Elander and Wilding, 1993	Self-report	Car drivers	711	3/1	Pearson	.305	Not reported	Also reported in West, Elander and French, 1992
Gebers and Peck, 2003	State records	Mainly car drivers	76,194	3/3	Beta	0.054	0.170 for first period	–
Häkkinen, 1979	Company records	Bus drivers	66	8/2.5–18.5	Pearson?	.56	25.26	Variable time periods between drivers
Johnson, 1938	State records	Mainly car drivers	30,000	6	Odds	(2)	Not reported	Odds of having another accident in the second time period, given at least one in the first
Krauss, Krumholz, Carter, Li and Kaplan, 1999	Self-report	Mainly car drivers	50 + 50	Variable	Odds ratio	0.465 (few previous crashes)	Not reported	Epileptic drivers. Case-control
Liddell, 1982	State records	Mainly car drivers	347 + 347	3.25	Odds	(2.05)	Not reported	Case-control, with cases being drivers injured in an accident. The odds of another accident within the same time period
Miller and Schuster, 1983	State records+self-reports	Car drivers	1,202 1,081 214	3/5 4/3 3/5	Pearson	.129 .090 .144	Not reported	The predicted time periods were 10 to 18 years removed from the predictor
Norris, Matthews and Riad, 2000	Self-report	Mainly car drivers	504	3/4	chi2	7.97 (previous MVA/ no MVA)	>0.46	Uncertain reporting. Second time period might have included near misses
Owsley et al, 1998	State records	Probably car drivers only	294	5/3	Relative risk	2.0	>0.639	–
Peck, 1993	State records	Mainly car drivers	161,303	Variable?/3	Pearson	.096	Not reported	The first time period appeared to be variable
Peck and Gebers, 1992 (as reported in Peck, 1993)	State records	Mainly car drivers	Not reported	6/5 8/3 3/3	?	.216 .170 .168	Not reported	Some uncertainty about the analysis. Culpable accidents 137, 128, 133

Note: Shown are the data sources, type and number of subjects, time periods used for comparisons, type of statistical analysis, results for all accidents and the mean of accidents in the population, and results and mean for culpable accidents only (if applicable). The time periods were mostly consecutive. Question marks denote uncertain reporting.

Table 3.3   The correlations between the (squared) size of the correlations between time periods from Table 3.1 and time periods and the mean number of accidents in each sample (total time period)

Variable	N	First time period	Second time period	Total time period	Mean no of accidents
Squared correlation size (all values)	78	.15	.25*	.20	.85***
Squared correlation size (outlier excluded)	77	.09	.19	.15	.93***
Squared correlation size, car drivers	30	.75***	.82***	.83***	.54**
Squared correlation size, bus drivers (outlier excluded)	40	.55***	.60***	.58***	.88***
Squared correlation size, tram drivers	7	.98***	.98***	.98***	.94***
Squared correlation size (outlier excluded, independent values only)	36	.18	.45**	.34*	.93***
Squared correlation size, culpable accidents	27	.56**	.70***	.65***	.80***

* p<.05, ** p<.01, *** p<.001.

Note: The first calculation thereafter subdivided into similar ones without outliers, for independent values only, and for three different driver groups.

studies with multiple, overlapping samples/values[12] resulted in the same result for mean number of accidents, and a strong increase in the association for the time period variables (sixth row, Table 3.3).

The regression equations of some of the correlations in Table 3.3 are also of some interest. In Table 3.4 these associations are shown, along with their mean

---

12   All values were included from Bach, Bickel and Biehl, 1975; Burg, 1970; Creswell and Froggatt, 1963; Daigneault, Joly and Frigon, 2002; Farmer and Chambers, 1939; Forbes, 1939; Harrington, 1972; Hauer, Persaud, Smiley and Duncan, 1991; McKenna, Duncan and Brown, 1986; Milosevic and Vucinic, 1975; Peck and Kuan, 1983; Stewart and Campbell, 1972; but only one (of several) from Blasco, Prieto and Cornejo, 2003 (the other value was deleted as an outlier anyway); Gebers, 2003; Gebers and Peck, 1994; Häkkinen, 1958; Moffie and Alexander, 1953; Peck, 1993; West, Elander and French, 1992; af Wåhlberg and Dorn, submitted; two from Peck, McBride and Coppin, 1971; and five from af Wåhlberg and Dorn, 2009a. Deleted were therefore mainly the data from the California Driver Record Study and Häkkinen. The principle was to use the sample with the largest N from studies with overlapping samples, and if similar, the longest time period was chosen.

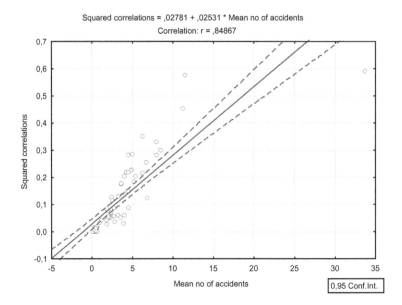

Squared correlations = ,02781 + ,02531 * Mean no of accidents
Correlation: r = ,84867

Figure 3.1   **The association between mean number of accidents in each sample for the whole time period used and the (squared) correlation of accidents between parts of this period (usually split-half), from studies in Table 3.1. N=78**

number of accidents. It can be seen that all starting points are close to zero, and that the increases in (squared) between time periods correlations are fairly similar for bus- and tram-drivers, while it is lower for car drivers. This could be interpreted as an effect of longitudinal changes in driver behaviour; car drivers can simply not achieve the same strong stability as high-risk populations, because when they have accumulated the same number of accidents, so much time has passed that the individual's environment has had a marked influence. This effect could also be due to the professional drivers having more homogenous driving environments. These interpretations lead to the prediction that for professional drivers, the stability of crashes in their private driving will always be lower when compared to their at-work driving.

Studies Not Included in the Meta-Analysis

Many unsafe drivers do not fall in the accident-repeater classification. McGuire, 1956b, p. 1742.

Table 3.4   The increase in between time periods correlations with time period and mean number of accidents for three driver groups

Population	N	Mean accidents/year	Equation for years	r	Equation for mean accidents	r
Car	30	0.063	-0.001 + 0.002 * years	.83	0.002 + 0.019 * Mean	.54
Bus	40	1.061	-0.004 + 0.039 * years	.58	-0.028 + 0.045 * Mean	.88
Tram	7	1.321	-0.055 + 0.063 * years	.98	-0.023 + 0.042 * Mean	.94

Note: The three different driver populations identified in the meta-analysis, their Ns (number of between time periods correlations) and mean number of accidents per year in the available samples. Also shown are the correlations between size of (squared) intercorrelations between number of accidents in different years and the total number of years used and the total mean number of accidents (=column three times the correlate in column four), and the regression equations for these correlations. The bus driver results were calculated with the Blasco et al. outlier removed.

A few studies on accident record stability over time were not included in the meta-analysis, for several different reasons. These studies are shown in Table 3.2. For most of these, the type of statistic used was not compatible with those in the meta-analysis. The reasons for excluding others were incomplete data, uncertain reporting and/or unusual methods used. Some of these will be commented upon here.

French, West, Elander and Wilding (1993) were some of the very few researchers to use exclusively self-reported accident data, and the only study to report a correlation (but see further on about a duplicative publication). For some reason, their resulting correlation was very much higher compared with other car driver data for similar time periods, including those where official registers had been complemented with self-reports (Miller and Schuster, 1983, three samples). The reason for this discrepancy is of course impossible to determine, as a simple difference in mean, due to higher reporting, should have had an effect for Miller and Schuster too (none of these articles reported their means). On the other hand, if self-reported accident data tends to be more reflective of drivers' reporting tendency than the real number, then a strong correlation between time periods would be found for such data, whereas when they are combined with official registers, the effect of the (probably) increased variance would tend to be cancelled out by the discrepancy between the sources, and yield similar values as official sources alone, as found for Miller and Schuster. However, the reason for the lower values in the last study could also be that their time periods were not consecutive. A further oddity about the French et al. finding is that it seems to have been reported in West, Elander and French (1992) too, (and therefore not included in the table) as part of a multi-wave questionnaire study. While the intercorrelation between the first time periods was extremely high, as noted, the second (between second and third time period) was not, and fitted perfectly into the meta-analysis correlation/ mean association. As no other comparable data have been published, this remains an anomaly, which could have important methodological and theoretical aspects. However, a simple error of analysis or transcription is of course a very possible explanation too.

Some other studies have also studied stability of accident record, but have not been included in the meta-analysis, or Table 3.2, for various reasons; mainly incomplete and uncertain reporting. Thus, Lefeve, Billion and Cross (1956) reported that drivers with no accidents in the last four year period had 0.8 accidents in total, while those with at least one accident had 1.8. The categorization of drivers was based on records, while the total number of accidents up to the study date was probably self-reported. If this indeed indicates any stability of accident record over time, it would seem to be very small.

Kahneman, Ben-Ishai and Lotan (1973) reported a chi2 value of 12.75 for the association of error severity scores for bus drivers' accidents in two successive one-year periods. It was not clear exactly on what data this value had been computed, and the error severity score is not the same thing as number of collisions (see Chapters 6 and 7).

Sichel (1965) found a correlation of .62 between accidents in different time periods for bus drivers, but the individual records had not been calculated as number per year, but as time periods between crashes, so they were not directly comparable with the data in the main meta-analysis of this chapter (this is discussed further in Chapter 6). Furthermore, Brown and Ghiselli (1948b) reported reliability coefficients (Spearman-Brown) of .32 and .42 for trolley car and motor coach operators. However, these values were calculated for odd against even months of an eighteen-month period, and it was uncertain whether the mean numbers of collisions of 5.18 and 7.76 were for the whole period.

Also, Hemmelgarn, Suissa, Huang, Boivin and Pinard (1997) found that drivers who had a crash where someone was injured were twice as likely as controls to have had a previous injury collision in the two years before the study period. Some indication of stability was also given by Rawson (1944), where drivers with four accidents in a three-year period had seven times as many accidents in the next period as those who had no crashes in the first period.

Maag, Vanasse, Dionne and Laberge-Nadeau (1997) reported that the association (Poisson regression coefficient) between taxi-drivers' accidents in two successive years was negative. With a mean of 0.5 accidents in the total time period, there should have been a positive correlation of about .16, so this result clearly does not fit in with those included in the main analysis. The reason for not including this paper, apart from the statistic, was that the years used were not the same for all drivers, something that could possibly have introduced seasonal variation into the analysis.

For motorcycle drivers, McDavid, Lohrmann and Lohrmann (1989) also reported some stability over time (for all types of recorded vehicle accidents). Only the longest time period (2/5 years) was significant, thus indicating an increase in association strength with time.

Finally, it can be noted that the cross-validation value of .109 (Φ coefficient) of actual versus predicted accidents for two three-year periods with a mean number of collisions of .341 found by Gebers and Peck (2003) would fit perfectly into the car driver meta-analysis regression line. However, it is not altogether clear how a predicted sample differs from the one the values were predicted from, so this data point was not included in the main analysis.

Overall, there is a number of studies that have reported results that do not fit in with those of the meta-analysis undertaken here. However, as these studies have not reported the details necessary for a full understanding of what might have caused these discrepancies, they remain anomalous.

Discussion

> ... the Connecticut Study ... concluded that once an accident record of a group is established as being predominantly accident-liable or accident-free its future history can be predicted with an astonishing degree of reliability. Tillman and Hobbes, 1949, p. 322.

Summarizing the results of the meta-analysis and various other data, it would seem that there is a very good case for claiming that a coefficient for traffic accident record stability is almost exclusively dependent upon the mean number and/or time period of collisions in the sample used, with type of population and unknown longitudinal factors having less impact. The length of the time period probably mainly has an effect by increasing the mean and variance, while over very long periods it introduces longitudinal effects that tend to counteract stability. If these results are extrapolated a little, it may be concluded that humans do indeed exhibit one of the features of accident proneness; a collision rate that is very stable over time.

The reason for the dismissal of the stability of accident record over time (and accident proneness) by previous researchers is apparently a misinterpretation of the weak associations between accidents in different periods in some studies. The problem is, however, statistical rather than psychological/behavioural; most of the studies presented in Table 3.1 used such short time periods that the disconcertingly low reliabilities found were inevitable. Contrary to the common view of a rather low stability of accident record, it may therefore now be concluded instead that accident record and accident-causing behaviour is very stable over time.

It is instructive to review one of the logical errors found in studies on crash record stability; Ross (1966), concluded about his group of accident-involved drivers that 'Their accident experience seems better than might be expected by chance ... This finding suggests a lack of support for the proposition that accidents tend to involve drivers with consistently poor driving performance.' (p. 24). Ross had found that drivers in crashes (they were all interviewed as a result of being in an accident) had very few collisions in roughly the two years preceding the last crash, as compared with the general population, and as a result he claimed not only that there was no stability of accident record, but also that drivers involved in crashes were in fact safer than others.

These remarkable conclusions were in fact based upon the error of not counting the accidents that were the reason for the drivers' inclusion in the study in the first place. If the total record of these drivers was included, they of course had vastly more accidents per driving year than the average (almost five times higher).

This faulty logic also illustrates the importance of the choice of time periods and (sub)-populations. In fact, the method used by Ross could not yield any useful data about accident record stability, because the time periods were different for different drivers, defined by their accident dates, thus introducing unusual exposure problems, as noted for similar variable time period methods in the introduction (see also Chapter 6). Also, the statistical properties of the sample became useless, as one of the time periods had no variability.

The pessimism emanating from many researchers about the stability of accident record, and thus the predictability of future collisions from previous crashes, has been fairly high. For example, Weber (1972) concluded, after having computed the confidence limits of his prediction of California drivers' accidents that 'From the standpoint of using accident records to discriminate between high and low predisposition to future accidents, the results in Table 5 are indeed disappointing.' (p. 114). This, however, would really seem to be unwarranted, unless one insists upon using very short time periods in low-accident populations. Similarly questionable is the statement of West, Elander and French (1992) regarding the intercorrelation between years in their data: 'This gives an estimate of the extent to which accident risk is consistent over time.' (p. 25). Instead, intercorrelations would seem mainly to mirror the limits of each study, as so far the upper limit of stability has not been ascertained. Neither can the interpretation of Smeed (1960) 'Consequently chance plays a large part in the actual distribution of accidents.' (p. 273) be validated by the present analysis. Actually, chance (in the opposite sense of stable individual influences, see further the discussion later on in this chapter) would rather seem to be the weaker contributor.

Limitations of the Meta-Analysis

The high-mean accident correlation of Blasco, Prieto and Cornejo (2003) did not follow the trend of the other data in the meta-analysis. It might also be noted that this result cannot be dismissed as a freak occurrence due to a small N, as massive amounts of data was available. In the end, no adequate explanation for this anomaly is forthcoming. As a further complication, the outlying value came from a study that supplied another data point for the same drivers, which fitted in well with other data. The only differences were the long period of observation and the very high mean of accidents for the outlier. This could be interpreted as a breakdown of the association found in the other data at longer time periods and/or high absolute numbers of crashes per driver, such that there is no longer an increase in the intercorrelations. However, although longitudinal changes in both drivers and environment could cause such an effect, such an interpretation is not supported by the results of Peck and Gebers (1992), which for similar lengths of time did not show any tendencies to level out. The increase in mean number of crashes of the population would therefore seem to be the reason for this discrepancy.

Two possible explanations can be suggested. Professional drivers with extreme numbers of accidents may be targeted by their employer for interventions that impact by lowering their accident rate. However, for such a process to have the result described it would need to be implemented after different (and fairly long) time periods in the Spanish company involved, about three or four years. Also, it could be the case that drivers react to their crashes and become safer with time. This could be tested by comparing the mean number of collisions of drivers in the second time period, categorized by whether they were over or under the mean in the first period. Similarly, the correlations of these groups could be calculated separately. However, none of these mechanisms seem to have been investigated.

It is therefore difficult to speculate about the upper limit of the increase in stability of the accident variable. About twelve years and twelve accidents would seem to be the limit of linear increase in the meta-analysis data, and beyond that, what happens is not known.

When evaluating the finding of a high correlation between the mean of a private driver population's accidents and the intercorrelation of accidents between time periods, it should, as noted, be borne in mind that much of the data come from official records, meaning that they are seriously under-reported (for example, Smith, 1966; Bull and Roberts, 1973; Harris, 1990; Aptel, Salmi, Masson, Bourde, Henrion and Erny, 1999). It could therefore be expected that the actual correlation for a certain time period is much higher, due to the higher mean. However, such a conclusion is built upon the assumption that minor and major accidents are fairly highly correlated within individuals (as severe accidents are more likely to be reported into official records than trivial ones (Smith, 1966), that is people are not differentially disposed to having very severe or very trivial mishaps, and as seen in the section about 'stability of accident record over environments' (stability between different types of accident), there is very little data available concerning this point. However, as the professional driver data usually contains all levels of severity in their accident counts, and these seem to have a slightly superior stability, it may be concluded that the real stability for car drivers is at least not lower than the presented figures.

Almost all private driver data in the meta-analysis was North American, which limits the conclusions somewhat, whereas the professional driver data was from Finland, Germany, Ireland, Spain, Sweden, United Kingdom and Yugoslavia. As the latter still fitted fairly well in with the car driver data when the different group means were taken into account, the question of whether the findings are representative should not arise.

Consequences

Methodologically, it is interesting to note that one of the consequences of finding good stability of accident record over time and between populations is that researchers do not really need to calculate such intercorrelations, as they may be approximated with a very high degree of certainty from the mean, at least for professional populations. This means that the reliability of the accident variable can be estimated, and a suitable time period for accident data chosen. These results also underscore the need for researchers to increase the length of time periods used for gathering accident data. The commonly used three-year period (af Wåhlberg, 2003a) is far too short, especially for car driver populations.

A few weaknesses and interesting blind spots may be noted in the literature and thus this meta-analysis. The dearth of stability data from self-reported accident studies has already been mentioned. However, even if such studies were undertaken, it would be hard to interpret the results, whatever they were, because memory and common method variance effects would work against each other, and there is no way of telling which mechanism would get the upper hand.

Among the professional drivers, there were only two sub-groups well represented (bus and tram drivers), while there was only one data point for truck drivers. Many other categories do not seem to have been studied in this respect, for example, taxi drivers, which could be assumed to drive under very different circumstances as compared with other professionals. Similarly, motorcyclists have hardly been studied at all regarding accident record stability.

Concluding Remarks

The notion of accident proneness has been a matter of controversy within the safety research community for decades, but apparently, the critics have won out in the end, and precious few seem to defend, investigate or even care about this matter, in whatever guise it is presented. However, at the same time, accident proneness would seem to be a widely accepted assumption when considering traffic safety research designs, although not by that name. All studies that have tried to predict individual differences in accident record would be meaningless if there was not some sort of personal factor (the predictor) that was causing the mishaps. This is especially so for the studies using longer time periods for their accident criterion. A vivid example of the rejection of accident proneness as a term and the concurrent acceptance of individual differences of (stable) accident-causing characteristics can be found in Crawford (1971), where the author stated 'It is now becoming clear that I do not subscribe to the philosophy of the accident-prone concept ...' (p. 907), but also 'The impetuous, selfish, inconsiderate, conceited man is likely to have accidents.' (p. 908). Throughout that text, there are statements about individual characteristics that pre-dispose their bearers to accidents, while there is a wholesale critique of accident proneness. How this argumentation was found to be logical by the author (and the reviewers) is inexplicable. The only solution would seem to be some sort of very extreme definition of the accident proneness concept, a common rhetorical trick used to make opposing views seem ridiculous. Yet, Crawford did state that 'Accident proneness ... is an abstraction in psychology which refers to a personal, stable and enduring characteristic that predisposes an individual toward having accidents.' (p. 905). How this definition differed from the statements by the same author about what causes individual differences in accident record is not clear.

This somewhat extreme example is instructional in many ways, because it brings out exactly the double standards that can be found in the (traffic) safety research as a collective; while there seems to be widespread reluctance to accept that there are predisposing factors, there would also seem to exist a general gut feeling that this is very much so. Why this is the case is very hard to discern.

Summing up, there is good reason for the traffic safety community to revisit the issue of accident proneness, and officially accept the reality of the strong stability over time of accident record. However, the type of analysis presented here is virtually unprecedented,[13] and certainly in need of further development and corroboration. Similarly, there is room for theoretical development; the possibilities of the accident proneness concept are far from fully explored. For example, research (not to mention theory) concerning the effects of culpability on stability is rare, while that of exposure is virtually non-existing. These two factors will therefore be analysed in the next two sections.

Culpability as a Stability-Enhancing Categorization

> Prior accident involvements are an important factor in estimating future accident risk; however, models using culpable accidents do not perform as well as models using total accidents. Gebers, 1999, p. vii.

Traditionally, all accidents of a certain broad type (industrial, home, traffic etc) have been lumped together for stability calculations, while few researchers have tried other categorizations implicit in accident proneness theory; associations between minor and major incidents, for example (among the few exceptions are Adelstein, 1952). However, the notion of culpability, or responsibility for the accident in terms of causing it, does not seem to have been a part of this line of thinking. This is somewhat strange, as situations can easily be imagined where persons are injured by the actions of someone else, without having the slightest possibility of protecting themselves, and much less having a hand in the causation of the event. Such an incident would rather be an effect of the other person's

---

13   The statistically minded reader may also consult the assumption- and distribution-filled paper by Holroyd (1992), where some analyses resemble the ones presented here, without reaching any similar conclusions.

accident proneness, and cannot contribute to the stability of the injured person's accident record, unless meeting up with clumsy people is a feature of this person's life. Also, (differences in) a stable rate of exposure to a risky environment might create a measure of stability for non-culpable accidents. All in all, however, the restriction of stability calculations to culpable accidents should only yield an increase in association strength, when mean number of accidents has been held constant. The last point is something that previous researchers have not taken into account.

In Tables 3.1–3.2 it can be seen which of the stability studies that have used culpability as an alternative grouping. None of these studies have found this concept to yield a better reliability than 'All accidents'. However, several have noted that this was probably due to less variation, as the comparison was always against 'All accidents' in the same population, and the mean and variance therefore by definition was lower. As should be evident from the previous analysis, this is indeed an unfair way of comparing, although of course each researcher did not have much of a choice.

However, calculations could be run for these data too, similar to those of the main meta-analysis, with twenty-five cases of stability coefficients for culpable accidents were there was also data for 'All accidents' (that is a sub sample of the main meta-analysis). The results are shown in Table 3.5 and Figure 3.2.

Correlation-wise it appears there were some differences, but these were probably due to the much larger variance in the initial correlations, where 'All accidents' had twice the standard deviation of culpable incidents, across studies. However, the main question is whether the stability of the accident variable increases faster with culpable or 'All accidents'. Therefore, the regression equations of each correlation were scrutinized. For 'All accidents' (for bus drivers) this was $0.00407 + 0.04144 \times$ mean, and for culpable $-0.02420 + 0.06218 \times$ mean, which shows that although 'All accidents' have an initial upper hand, at a mean of 1.362 accidents they will be equal, and thereafter culpable accidents will have a higher stability.

Table 3.5   The correlations between squared between time periods correlations and time periods and accident means

Variable	N	First time period	Second time period	Total time period	Mean no of accidents
Squared correlation size, culpable accidents	25	.48*	.65***	.58**	.77***
Squared correlation size, all accidents	25	.82***	.82***	.84***	.92***

* p<.05, ** p<.01, *** p<.001.

Note: Shown are correlations between correlation sizes for culpable and all accidents versus the time periods used and the accident means of the samples with results for culpable accidents in Table 3.1.

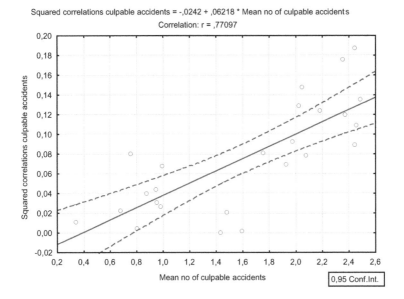

Squared correlations culpable accidents = -,0242 + ,06218 * Mean no of culpable accidents
Correlation: r = ,77097

Figure 3.2 **The association between mean number of accidents and the (squared) correlation between accidents in different time periods in the samples, for culpable accidents only, from studies in Table 3.1. The slightly outlying values in the lower middle were from Häkkinen (1958). N=25**

Thus, the common assertion that culpable accidents have a worse reliability as compared to all (for example, Peck, 1993) does not receive any backing when a fair comparison is made. If anything, the opposite would seem to be likely, although the small amount of data and overlap between samples make these calculations quite provisional.

One other interesting possible calculation could be made on data from Table 3.1. In Gebers and Peck (1994) and Gebers (2003), responsible accidents were used to predict 'All accidents' in ten (overlapping) samples. As this is different from the principle described above of predicting culpable from culpable accidents, these data were not included in the previous analysis, but as there is also a certain overlap, these ten cases were run separately. Here, total number of years correlated .98 with squared correlation size, while mean number of accidents did worse; r was only .90.

Furthermore, Chandraratna, Stamatiadis and Stromberg (2006), found that the culpability of drivers in crashes could be predicted by their previous at-fault accident involvement. The odds ratio was only 1.17, but it is uncertain how the peculiarities of the sample (only crash-involved drivers, probably erroneously categorized culpable/non-culpable) might have influenced the results.

One more study deserves being mentioned here; Chen, Cooper and Pinili (1995) used culpable accidents only, finding that previous such crashes were the best predictor of later ones, but unfortunately did not report any correlation coefficient.

Finally, it should be noted that how culpability has been conceptualized and assigned in these studies has probably had some impact, a problem that is discussed in Chapter 4. In general, it can be suspected that the values presented here are under-estimates of the real associations, due to erroneous coding of culpability.

Exposure as a Stability-Enhancing Factor

> If one waits long enough and keeps an accurate record of accidents it becomes clear, after a number of years, that certain individuals have a sufficiently high accident record to make it unsafe for them to continue driving. Tillman and Hobbes, 1949, p. 330.

The stability calculations presented in Table 3.1 have, almost unanimously, one feature in common; the raw values were not controlled for exposure per time period.[14] This is somewhat disturbing, as amount of exposure regularly has been found to correlate with accidents,[15] although the association is usually weak (see further Chapter 5). Therefore, at least a modest increase in stability could be expected if exposure was held constant. However, in af Wåhlberg and Dorn (submitted) this had a slight positive effect for 'All accidents', but negative for culpable ones, which is peculiar. Meanwhile, it can of course be concluded that exposure has little impact in populations which are fairly homogenous for this variable. Also, the difference in regression lines between private and professional drivers in the present correlation/mean by population calculations would seem to point to a somewhat higher initial reliability for the professionals, which in its turn can be suspected to be due to a more homogenous exposure to risk, both concerning amount and levels.

In summary, little is known about the effects of exposure on the stability of accident record, although this is a promising avenue of research, which was pointed out long ago (Crawford, 1960), and many researchers take exposure into account when working with accidents in other areas. However, there are reasons to believe that the real association between crashes and exposure is far from understood, and that future research may therefore be able to shed some light on how exposure

---

14   This seem only to have been done by Liddell (1982) and Krauss, Krumholz, Carter, Li and Kaplan, (1999) by case-control methods (see Table 3.2); and af Wåhlberg and Dorn (submitted), using hours of work as measure of exposure.

15   For example, Greenshields and Platt, 1967; McGuire, 1972; Jonah, 1990; Dionne, Desjardins, Laberge-Nadeau and Maag, 1995; Lawton, Parker, Stradling and Manstead, 1997b.

influences stability. Now, we turn to another feature of the current thinking about the concept of accident proneness; whether people with many accidents of one type also have many of other types.

Stability of Accident Record Over Environments

... involvement in one kind of accident is quite independent from involvement in other kinds of accidents. Salminen and Heiskanen, 1997, p. 35.

If accident proneness is a general trait, then a person who has many driving mishaps should have all kinds of them, as well as all other sorts of non-driving accidents, that is accidents in different environments should correlate. Here, research is less prolific, and so no meta-calculations can be undertaken. Instead, what results have been found will be briefly described.

Starting with associations between different types of traffic accidents, Brown and Ghiselli (1948a) found that collision and non-collision accidents correlated .25 for motor coach operators. With a mean of 0.763 accidents per month, this value is a bit low, even considering the short time period used (maximum 20 months). As seen in Chapter 1 (see also Chapter 7), the evidence concerning the correlations between major and minor accidents is scarce and inconclusive.

For road accidents at work and in private driving, only two studies seem to have been reported; Newnam, Griffin and Mason (2008), who found an association of .06 for a six month period, and af Wåhlberg (2002a), where bus drivers' collisions correlated .24 (three years) with car crashes. Similarly, Albery, Strang, Gossop and Griffiths (2000) found a correlation of -.69 between crashes where the driver was impaired and not (the time period was probably variable). However, all these values were for self-reported data.

For traffic accidents versus other types, Wallace and Vodanovich (2003) had a correlation of .69 between work and automobile accidents. However, these data were also self-reported, which means that the association is probably inflated (af Wåhlberg, 2002a). Lefeve, Billion and Cross (1956) reported that people with recorded traffic accidents during the last four years tended to report more accidents in other environments. Unfortunately, no correlation was given, and as the traffic accident data was dichotomised, there was little use in calculating an association measure from the data presented in the paper.

Concerning correlations between traffic accidents with different types of vehicles, there is really very little research. Davis and Coiley (1959) reported that many of the drivers who had a record of safe driving with their firms had several accidents according to police records. However, no calculations were presented. As reported above, in af Wåhlberg (2002a), a correlation of .24 was found between bus and car crashes. However, these were self-reported data; between self-reported car and company recorded bus accidents the correlation was -.04 (ns).

Salminen and Heiskanen (1997) used interviews from fairly large Finnish samples to see whether accidents at home, traffic, work and sports were associated, but the results showed very weak correlations (the highest being .054). Similar results were reported by Salminen (2005), using the same method. Unfortunately, means were not reported in these studies. However, it can be suspected that accidents were very few, and the results therefore not really indicative of weak associations. Guilford (1973) on the other hand, found correlations ranging from negative to .31 between different accident measures, including observed incidents in a mobile kitchen under controlled circumstances. Here, traffic accidents correlated .08 to .20 with five other variables, despite a mean of 0.20. However, the means of the other variables were about ten times higher, so this probably explains these effects.

Also, Kunkle (1946) found correlations ranging from .06 to .32 between pilots' flying accidents and other types of mishaps, including such innovative measures as the number of scars on their bodies (.32). However, this was a contrasted sample, so coefficients were probably larger than for a random group. Also, these were military flyers straight from the war, which might also have had some sort of impact, like selection of who had survived.

Given that various correlations, weak though they are, seem to exist between different types of accidents, it would be feasible to do a factor analysis of such mishaps, to see whether there is in fact one accident proneness factor, or several dimensions. However, only one such paper has been found; Keehn (1959). In this study, it was tentatively concluded that there seem to be three different types of mishaps that go together, due to how the body of the person was involved (injuries to hands when using them, injuries to other extremities without using them, gross bodily movements resulting in injuries). However, as the data was self-reported, a real effect and common method variance could both explain these results. What can be concluded is that this type of analysis could be very worthwhile for furthering our understanding of accident proneness.

Leaving traffic accidents behind, Adelstein (1952) found no correlation between home and industrial accidents for five years, but .1867 for an eleven-year period. Furthermore, major and minor injuries correlated only .102 (ns) for five years, and property damage and injury accidents only .028 for four years.

In conclusion, there seems to be very little research concerning the stability of accident record between areas, and what little has been reported is often very poorly described regarding the methods used. None of the reports studied seem to have taken culpability into account. It is therefore difficult to draw any conclusions at all whether the evidence supports or refutes the notion of accident proneness across environments.

Stability Over Time of Other Types of Accident

... the repeater group changes its composition from year to year ... Williams, Henderson and Mills, 1974, p. 99.

Concerning other types of accidents rather than road traffic crashes, only a few studies have been located that have reported values on between time periods associations, which can be compared with the accident means of these samples and the time periods used. Also, the environments have been somewhat differing, and it would therefore be a precarious undertaking to add these values together in a single meta-analysis. However, as these studies have often reported several values, a small analysis could be calculated within each.

In Table 3.6 data can be seen from the very first study on accident proneness by Greenwood and Woods (1919). A dependency of the between time periods correlation would seem to be evident in these data. Similarly, Adelstein (1952) found that shunters' accidents correlated between time periods (see Table 3.7). It can be seen from the table that they did so in a way very similar to traffic accidents (the correlations between size of correlation on one hand and mean and total number of years were both above .60). Furthermore, property damage accidents also correlated as expected with their mean (Table 3.8), and if these were added to the data in Table 3.7, the r for correlation versus mean rose to .79 (N=7, p<.05). This was also found for data given in Newbold (1927), where the correlations of a dozen industrial workers correlated .30 with time periods used.

Data from Whitlock, Clouse and Spencer (1963) is shown in Figure 3.3, but these authors also calculated a correlation of .37 between factory worker injuries during a total of three years, but only .06 between four month periods, while Wong and Hobbs (1949) found that the workers with the highest accident record in a four-week period also had 2.5 times as many accidents in the next period as those with few accidents in the first.

Table 3.6 **The correlations found between industrial workers' accidents in different time periods by Greenwood and Woods (1919), and the means of accidents in the total time periods**

Time period (years)	3/3	3/3	3/3	3/3
N	21	22	36	29
Correlation	.72	.37	.69	.53
Mean number of accidents, total time period	3.85	2.32	3.16	2.27

Table 3.7   **The correlations found between shunters' accidents in different time periods by Adelstein (1952), and the means of these accidents in the total time periods**

Time period (years)	2.5/2.5	2.5/2.5	3/2.5	3/3	6/5
N	182	122	122	122	122
Correlation	.138	.200	.0482	.109	.258
Mean number of accidents, total time period	1.32	0.98	1.16	1.26	2.25

Table 3.8   **The correlations found between shunters' mishaps (property damage accidents) in different time periods by Adelstein (1952), and the means of these accidents in the total time periods. N=154**

Time period (years)	1/1	1/1	1/1	2/2
Correlation	.193	.326	.309	.355
Mean number of accidents, total time period	1.53	2.00	.198	3.53

Boyle (1980) reported a Spearman correlation of .71 between occupational accidents in the halves of eight years and nine months of work for press shop workers. This study is noteworthy for its comprehensive listing of important methodological factors in accident proneness research, and also for its comparison of the accident distribution in the data to other distributions. It was found to be significantly different from a Poisson distribution, which could be expected for such a long time period.

A very high value for stability over time of injuries in self-reported data was published by Porter and Corlett (1989); .9 for two waves of a questionnaire, spaced eighteen months apart. Although no means were published, two pieces of information can help the interpretation of this extreme value; it was calculated for a summary score of answers to six items about very minor injuries, and contrasted groups were used. Both these mostly have a very high variability and mean as result. For some reason, the authors concluded that '... a *belief* in a propensity towards minor accidents is a stable trait.' (p. 330, italics added). Why they choose to interpret their data (self-reports about actual injuries) in this way is uncertain, as they did not present any argument or data to suggest that what they had measured was not a stability of behaviour but of beliefs.

Although the data reported in this section was considered rather diverse, as well as sometimes lacking accident means, the same trend as for the traffic accident

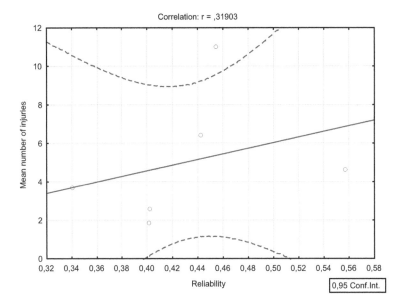

Figure 3.3   The reliabilities (Spearman-Brown corrected correlations) found for factory workers' accidents by Whitlock, Clouse and Spencer (1963), and the means of accidents in the total time periods. Data taken from table 1

record stability analysis was evident, with longer time periods and higher means yielding stronger correlations.

Discussion of Stability Within and Between Other Types of Accident

> The typically low correlation between the number of injuries in successive accident periods has been recognized by some writers as evidence of the low reliability of the number of injuries as a criterion for accident proneness ... Whitlock, Clouse and Spencer, 1963, p. 35.

The data for stability of accident proneness between accident types are fairly scarce. Unfortunately, this state of affairs is rather understandable, as often there is access to one source of data, but not another, unless it is self-reported. So far, the evidence seem to indicate that stability is very weak (as concluded by Ghiselli and Brown, 1955), but given the meta-analyses undertaken on other data in this chapter, such a conclusion is premature.

Conversely, it could be asked, why would an accident syndrome/proneness not exist? Or rather, why and how would a tendency to have traffic accidents, for example, be determined by a personal trait that had no impact on other areas of safety? If accident proneness is rejected, another explanation is needed as to how lack of individual safety in different areas relate. So far, no one seems to have constructed such a theory.

Another issue, which could be discussed as a result of the meta-analysis results, is whether humans learn from their mishaps. Having established that there is indeed a great deal of stability in drivers' accident record over time, it could be asked why this is so. If one does have an accident, will this not influence how you drive in the future, that is will you not become more cautious? Although such an effect would seem to be apparent for exposure, there is precious little research about how our behaviours relate to injuries and accidents (for example, Rajalin and Summala, 1997).

The Influence of Transitory Factors on Individual Accident Record

When researchers try to predict individual differences in crash record, the variables used can be, for example, personality, alcohol use or speed behaviour. All of these variables tend to be rather stable over time, and that is often a preference on the side of the researchers; we like our variables to show consistency over time, because that tends to increase their predictive power. However, crashes can also be caused by very sudden and unprecedented happenings, which can still be considered to be individual differences variables. These factors could be illnesses, stress and several other, which might be of very differing durations. Here, it is interesting to ponder how very short-lived factors that influence a driver's propensity to cause crashes relates to accident proneness.

The interest in transitory individual accident-causing factors has been considerably weaker than that vested on the stable ones. In af Wåhlberg (2003a), 100+ papers on accident prediction were reviewed, and less than ten could be said to be about transitory factors. However, this statement is somewhat dependent upon the definition of the term transitory. Should only environmental[16] phenomena that alter the driver's behaviour be considered? Or could a state of mind with little known external reason be included, for example, a schizophrenic episode? What if transitory factors are uncommon but recurring, like bouts of drinking and driving once a year? Suicide by automobile, however, is not considered in the present discussion, only unintentional events.

Some studies have found transitory variables to be important in accident causation, for example, quarrels and physical fights shortly before crash (Selzer

---

16   Here, the term 'environment' is (unfortunately) used in two senses; events or milieus which alter the state of mind and behaviour before or during driving, but which are not part of the driving environment proper, as they are not part of the traffic system. The other is the driving environment in the sense of roads, other road users, vehicles, traffic control systems etc.

and Vinokur, 1975; Selzer, Rogers and Kern, 1968; Holt, 1982). Similarly, people involved in divorce proceedings have been found to have an elevated risk of accident (McMurray, 1970).[17] However, these drivers did have a higher accident rate long before their marital break-up. Regarding depression, Isherwood, Adam and Hornblower (1982) did not find any differences between a group of accident-involved drivers and a control group. Similarly, Armstrong and Whitlock (1980) found no differences between psychiatric and physically ill patients, although heavy drinkers had higher rates. This was also found for alcoholics by Payne and Selzer (1962). Also, various 'life stressors' have been linked to accidents, for example, on the job (Norris, Matthews and Riad, 2000), as well as peptic ulcer (Smart and Schmidt, 1962), which could be an outcome of stress.

However, a common drawback to these studies (given the present perspective) are that they do not control for commonness of the explanatory factor, or stated in another way; are these transitory factors only examples of a wider syndrome of actions for these persons? The man who has quarrelled with his wife and has an accident the next hour, is this due to the quarrel, or is it due to the man's personality (or socio-economic status, or whatever type of explanation is preferred), which causes quarrels to be common in his family? Or could it even be said that the cause is the personality trait that makes him drive off in anger, instead of having a walk in the woods?

If no transient factors effected accident liability in an individual's life, crashes would probably be fairly randomly distributed over time. If, on the other hand, they did have an effect, this should show up as aggregations in time in a person's accident record. This kind of thinking has been tested by a few researchers, with some fairly surprising results (given the low interest shown by most traffic psychologists and traffic researchers in general).

The study by Blasco, Prieto and Cornejo (2003), included in the stability calculations described above, did not have the primary goal of showing how stable drivers' accidents records are over time, but rather the opposite. In a population of Spanish bus drivers they found that accidents tended to occur in short time intervals, despite the high inter-correlations between time periods. Similar results were presented by Creswell and Froggatt (1963). This observation might be an important clue for the interpretation of the accident/individual phenomenon under discussion.

A similar but slightly different line of thinking argues that it is not any transient factors in a person's life which cause sudden bursts of accidents, but that the experience of a first crash influences the probability of having another. Thus, Horn (1947) found that, for aircraft pilots, the time intervals between accidents were much shorter than could be expected on a 'chance' basis (that is what would here be called a very stable individual liability to have accidents). However, rather much the opposite was found by Mintz (1954b) for cab drivers.

---

17   The methodology of these four studies was of low quality.

In summary, it could be said that there is a case for the effect of transient factors on accident liability (regardless of whether these are previous accidents or other factors in the driver's life). The question is how much this factor limits the stability of accident proneness. No research on this question has been located.

Discussion

> By definition, an accident is a happening that is determined by chance. Thus, the frequency of accidents is determined by the basic risk inherent in any situation. Tillmann and Hobbs, 1949, p. 321.

The present chapter has investigated the evidence for stable and transitory individual factors as determinants of accident record, mainly regarding traffic accidents. Overall, the data would seem to indicate a much stronger stability of traffic accident record over time than has ever been suspected before, and it has been shown that previous conclusions (for example, Evans, 2004) have been based upon a statistical/logical error. For other types of stability, the evidence is less conclusive, mainly due to a dearth of studies. Having established this, we need to discuss how this evidence is to be evaluated in relation to the accident proneness concept, and if there is any other theoretical construct which could explain the present findings.

One of the arguments against accident proneness theory forwarded by its opponents (for example, Froggatt and Smiley, 1964) is that it is virtually impossible to test the proneness part separately from the stability of the exposure to risk. In principle, each person could have a certain environment that is stable over time, which creates the stability over time of their accident record, while the individual component is non-existing. However, there are some data that contradict this argument. The driving environment for bus drivers in the study af Wåhlberg and Dorn (submitted) was extremely homogenous, with all routes passing through the centre of town, and similar suburban areas on the outskirts, and all drivers driving several different routes, at differing times of day. It could also be noted that for these drivers, the driving style in terms of accelerations was found to be an individually stable variable for several years, that differed very little with the influence of traffic volume (af Wåhlberg, 2003b; 2007a; 2007b). Despite this homogeneity of the environment, the between time periods correlations of these drivers fitted perfectly in the present meta-analysis. If the influence of the environment is strong, then the coefficients for this population should have been much stronger than the others, in relation to the means in the samples.

The risk exposure argument also does not fit in well with the findings of the meta-analysis of culpable accidents. In fact, if it was the environment that was causing the stability, non-culpable accidents (and therefore 'all') would be equally stable to culpable accidents. Apart from the initial difference at the very low means, this does not occur. If non-culpable accidents had a stability of a similar strength as to culpable ones, there would be no difference for these regression lines (described in the culpability section above). Furthermore, in af Wåhlberg and Dorn (submitted), the stability of non-culpable accidents was calculated separately, and found to be about one tenth of that of culpable ones for the same time period. Despite non-culpable crashes having only half of the mean of culpable ones in this study, this difference is fairly large.

However, the unequal exposure argument is in many ways only an assumption. Those who forward it as an argument against accident proneness really need to show that it actually has some stabilizing effect, which they usually do not (for example, Froggatt and Smiley, 1964). This kind of argument can be advanced against any research (that is an unknown but possible bias has contaminated the results), and therefore need some corroboration. Actually, Häkkinen (1958) calculated the differential risk between different routes for bus drivers, and how much this would possibly influence their accident record. He found the effect negligible. In the end, even if there is an unequal exposure effect, this is probably much smaller than the individual differences between drivers.

Yet another criticism of accident proneness is referred to as the reporting tendency (Sass and Crook, 1981). When working with distributions, it is apparently problematic not to know whether the data is a 'true' representation, or some sort of truncated version of the actual distribution, due to under-reporting. In stability measurements, this is probably not a problem. If there is some sort of reporting bias, that is minor accidents go unreported, the correlations between time periods will simply shrink. This can be tested by removing parts of the data in steps in a stability calculation. If there is a systematic bias of reporting between drivers, in terms of a correlation between reporting tendency and number of accidents, the stability will be under-estimated, regardless of whether the correlation is positive or negative. Both will lead to less variation in the sample.

One of the problems concerning the lack of acceptance for the accident proneness concept would seem to be the very strong formulation of the concept by the opponents. For example, Froggatt and Smiley (1964) said: 'If accident proneness were a stable entity, as classically conceived ... the same individuals should be present in the 'tail' of the distribution in every observational period of reasonable duration.' (p. 6). This formulation in itself contains two requirements that are virtually impossible to achieve simultaneously; the distribution should have a tail, meaning being Poisson-distributed, and the drivers should be the same. This exactness simply cannot be found for such a short time period/skewed distribution as a Poisson. One could just as well demand that the same individuals should have the same number of accidents every day. This argument is therefore simply the old rhetorical trick of re-stating your opponent's view in an impossibly strong way.

A possible model for accident-causing behaviour, which would integrate the available data, is that a driver has an inherent accident liability (or proneness) that predisposes him/her towards having a certain mean level of incidents. However, this is not a totally stable trait, but could rather be seen as a mean around which the values fluctuate. This may be more apparent if accidents are replaced by dangerous behaviour (whether intentional or not). Some people are basically very cautious, others reckless, but no one is so all the time. We all have ups and downs, as well as long-term changes, and when another environment is encountered (with a different inherent level of risk), this mean level with its variation will just be transposed a notch. A cautious person will have few accidents in a fairly safe milieu, and more in a dangerous one, but relatively fewer than the reckless person within each.

In summary, there are not really any contradictions between the data concerning stability versus fluctuations, if these are compared over very long time periods. In other words, the differences between these somewhat conflicting views are probably only in existence when looking at rather short time periods, say less than ten years.

This line of thinking can be illustrated by the case of a single bus operator (see Figure 3.4). This driver has been working for the same company for close to two decades, and has one of the worst safety records of this employer (all the rivals to this dubious honour have quit after a few years of service). However, despite being a notoriously dangerous driver, there are some years when no incidents have been reported for this driver, and the accident record, if tabulated by single years, is very uneven. Two things may be noted regarding this; first, the importance of knowing the exposure when calculating stability coefficients. Part of the reason for this driver having no accidents in 1994 was probably because his licence had been temporarily revoked. However, the incident-free years of 2001–2002 he was working full time. Thus, despite his bad record, he 'suddenly' managed to drive safely. Second, longitudinal effects may be involved, both the driver's own aging and increased experience as well as a trend towards fewer accidents in the company as a whole (data before 1994 are not complete, so no calculations can be made to ascertain this hypothesis).

This example is a good illustration of the competing influences noted in the literature. There seem to be a basic dangerousness about this driver (most of the years he is a great deal above the mean for the population), while other, non-stable factors, would also seem to be at play, causing the inconsistency of his record.[18] Finally, there appears to be a long-term trend towards a lower accident rate. All these mechanisms make sense, and are well known. What may have been lacking so far is a more integrative approach, where all of the factors are recognized simultaneously.

At this point it is important to discuss the philosophical difference between the psychological approach taken here considering the causes of accidents, as compared with the factors inherent in the statistical distribution techniques. For some reason, many researchers who fit a theoretical distribution, for example, Poisson, to actual accident data, and find no difference, seem to conclude not only that there is little if any stability in the individual accident liability, but also that

---

18   It might be noted that this inconsistency may be an especially large problem for case-control studies that use short time periods (for example, Leveille, Buchner, Koepsell, McCloskey, Wolf and Wagner, 1994), although comparative studies are lacking.

All accidents per year of a single bus driver

Figure 3.4   The number of accidents (regardless of culpability) per year experienced by one bus driver

the empirical distribution is 'random' (for example, Adelstein, 1952) and that accidents are therefore due to 'chance' (for example, Smeed, 1960). Here the discussion usually stops, implying that researchers in this vein look upon accidents very much in the way of the old adage 'one time is no time, two is a habit', that is if a human has repeated accidents, these are due to behaviour, but a single accident is 'random' and thus not due to the behaviour of the individual.

Such a view is not accepted here, for several reasons. First, saying that an incident is 'random' is no explanation at all as to why it happened, and is therefore unsatisfactory from both a scientific and practical perspective. Second, there is always a cause to an accident, and even if a human has only one in a lifetime, this one may still be very much due to the individual's behaviour, and therefore in principle possible to predict. This difference in view would seem to be between looking upon mathematics and statistics as either a description of reality, or as a cause, and is apparent for example, in the comments by other statisticians included in the paper by Adelstein (1952).

In general, the findings presented here are only predicted by the accident proneness theory. No other theoretical formation seem to incorporate this aspect, although this would often seem to be due to oversight, and not because the concept of accident proneness is incompatible with other conceptions about safety and accidents. Rather, they are probably complimentary. However, a comprehensive treatment of accident theories is outside the scope of this text.

Some methodological consequences of accident record stability findings were discussed in the introduction; a variable time period between subjects for accident record is clearly not a valid method in studies of individual differences studies in safety. Furthermore, the strong dependence of the correlation size in the present meta-analyses on the mean/length of time period of the accident sample probably has the same effect in other prediction studies. This means that meta-analyses of effects will estimate the mean effect given a mean number of accidents in the criterion. This will probably result in an underestimation of the real effect. Therefore, reviews like those of Clarke and Robertson (2005) about the influence of personality on accident record are not really meaningful until the mean number of accidents is stated, that is a certain strength of association will only be valid given a certain amount of variance in the criterion.

It would also seem to be necessary to recommend the use of longer time periods for the calculation of accident criteria, as well as the use of high-risk populations for research. Otherwise, the statistical power of the study will simply be inadequate, which is often the case today.

The present results clearly indicate that traffic accident liability is very stable over time, and that the pessimistic view of many authors concerning the characteristics of this variable is unwarranted. However, the picture of accident stability is one thing, while the practical benefit of this finding is somewhat different. First, it can be stated that many researchers have let their negative views of accident stability influence their recommendations on its use. For example, Lawton and Parker (1998), claimed that '... accidents themselves are largely meaningless as outcome measures, being unreliable and stochastic in nature.' (p. 667). These authors instead recommended their own principle research variables, errors and violations, as dependent variables. Why these would be a better criterion, apart from being easier to measure (in terms of yielding data, but not in terms of validity, as they are always self-reported), is uncertain. The claims of validity for the DBQ rest upon the associations of the violation scale with accidents, and the recommendation is therefore illogical. If collisions cannot be trusted as criterion, then there is no evidence for the DBQ in the first place. As was shown in Chapter 2, the DBQ can also be questioned for several other reasons.

Instead, we are again reminded of the importance of the crash variable, as well as the disenchantment of many safety researchers with this parameter. However, being displeased with a variable because it does not yield the expected results is not really a good criterion for anything, and there is simply no valid alternative to collisions as measure of safety, as will be shown in Chapter 7. First, however, we return to the concept of culpability. As stated in the introduction of this chapter, the subject was really about whether the reasons for an individual's propensity towards causing accidents is a stable trait or reactions to the environment. The issue of culpability has therefore surfaced again, and it is high time to analyse this concept and the available research.

Chapter 4

The Determination of Fault in Collisions

... the judgement of probable cause or 'fault' of each accident from such information ... would be highly subjective. Stewart, 1957, p. 296.

Introduction

In previous chapters, the notion of culpability for traffic collisions has surfaced in several instances, and the importance of using this concept in traffic research has been emphasized. However, it has also been recognized that a practical problem exists, because culpability is hard to determine, not only for reasons of deficient information, but also because we do not really have an explicit, accepted and validated definition of responsibility[1] for accidents within the traffic safety research society. Similarly, there is no validated and accepted method for measuring relevant parameters or doing the actual categorization. In the end, each researcher and practitioner makes his or her own choice, and there has been no way of objectively investigating whether anyone is more or less right than anyone else about how collision involvement should be judged.

The lack of valid information is a methodological problem that can in principle be solved, given enough resources (the technological development of in-vehicle data recorders will probably help in the future). The main problem is instead that there is no generally accepted criterion for culpability for traffic accidents. Instead different researchers seem to use different criteria, and this probably leads to erroneous categorization of collisions, because the criteria themselves are ad hoc, without any empirical, statistical and/or theoretical foundation. In effect, it can be shown that many researchers, who have categorized according to responsibility, and especially those who have used the judgement of police officers, have used criteria that are too lenient, and therefore coded culpable accidents as non-culpable, introducing error into their calculations. The present chapter will investigate the evidence concerning criteria of culpability, discuss what testable consequences a correct criterion of culpability will lead to, and refer to the first research into the validity of criteria, and show that erroneous coding does exist.

It should be pointed out that in many instances, the guiding principle when assigning culpability has been that only one driver in a two-vehicle collision could be mainly responsible. This is explicitly stated within induced exposure techniques

---

1   In the present text, culpability is seen as being defined by whether a road user could in any reasonable way have avoided the collision, and still be present at the scene of the incident (af Wåhlberg, 2002b).

(see the next section), but does often seem to be the case in other circumstances too, for example, when police officers judge who was at fault.

However, the stance taken here is that for any collision, any road user involved may share the blame, in differing degrees. Therefore, while in a dichotomous at fault mode 50 per cent of drivers in two-vehicle crashes will be judged culpable, this figure will in practice be higher when blame can be shared. Adding practically all single collisions to this figure further increases the percentage of drivers involved in the overall collision involvement population who are to blame. However, before looking at the question of criteria of culpability, we will investigate how researchers usually handle this.

Induced Exposure, Responsibility Analysis and Case-Control

> It can be hypothesized that the probability of any driver being involved in a crash without having any responsibility is the same because an at-fault driver does not intentionally select whom to strike. Chandraratni, Stamatiadis and Stromberg, 2006, p. 535.

As was described in Chapter 1, some researchers within the individual differences approach to traffic safety have used culpability and similar concepts to categorize their collisions, trying to optimize their predictive power. Sometimes, this has worked out, sometimes not. What is apparent from the studies referenced in that chapter is that these researchers did not discuss the issue of culpability, or tested their assumptions, other than trying different sets of accidents as criteria (for example, Vernon, Diller, Cook, Reading, Suruda and Dean, 2002).

In contrast to this lack of interest can be held the two very similar approaches of induced exposure (Cerrelli, 1973) and responsibility analysis (Terhune, 1983), which are similar to the case-control method of medicine applied to traffic collisions (for example, Lardelli-Claret, Luna-del-Castillo, Jiménez-Moleon, Femia-Marzo, Moreno-Abril and Bueno-Cavanillas, 2002). Within these research methodologies, non-culpable collisions are used as a proxy for quantitative and qualitative exposure of driving risk (and in reality many other variables too), where such data is unavailable, which is most often the case for the general driving population. The logic is that a group of non-culpable drivers is a randomly selected sample from the population, and therefore have characteristics that are similar to those of the average driver (Cerrelli, 1973). By comparing the culpable and non-culpable drivers, differences can be found, which are thought to be causally related to their accident involvement.

Two important conceptual differences between induced exposure and responsibility analysis have been found (although not all studies using either of these terminologies adhere to them). Induced exposure often explicitly assumes that, in two-vehicle collisions, only one driver is to blame, while responsibility analysis allows for blame to be shared. Also, the latter method would seem to require the cases and controls to be independent, meaning not being in the same crash (Hours et al., 2008). Why this is required was not explained, and this requirement has not been mentioned by other authors using responsibility analysis.

In general, however, these approaches share one very important feature with the individual differences research where culpability has been used; no validation has been undertaken of the assumptions or the criteria. Testing the validity of induced exposure against other exposure measures is very uncommon (for exceptions, see Dorn and af Wåhlberg, 2008; af Wåhlberg, 2009), and other methods for validating the categorizing do not seem to be available.

Before we turn to the problem of validation, we will, however, take a further look at the results of research where categorization of culpability has been undertaken. The reason for this is that some systematic effects probably can be found regarding the sources of the judgements, which can be useful for future research. Also, it is important to point out how much these judgements differ between studies, which indicates a source of error. Where this error lies (too lenient or harsh a criterion, or differences between populations) is hard to know without testing in each set of data, but a guiding value will be given in the section below.

The Level of Culpable Collisions in a Population; Evidence and Methodology

> ... in every collision accident ... there will be one responsible driver and one that is not responsible ... Carr, 1969, p. 348.

Given a sample of a hundred drivers who have been in an accident, how many should we denote as culpable for their involvement? The question might at first seem strange, because is not this number a function of what we find when we investigate the collisions? Not really, because in each instance, we make a judgement and this judgement is based upon a criterion of what we consider to be a cause, or something similar. However, these criteria are not only implicit in most instances of the research that uses them, but also differ between studies. A harsh criterion will lead to a large percentage of culpable drivers, while a lenient one will have fewer. That such differences exist between judges is shown in Tables 4.1–4.4, where the percentages of culpable accidents (indicating the severity of the criteria) in some studies are shown in the second column. As can be seen, there is a fair amount of scatter, as the results range from 13 to 91.7 per cent. This means that different researchers use different criteria.[2]

---

2   In these studies, most have actually let someone else do the categorization for them, but it is still the researchers who use the data and accept how culpability has been assigned.

Table 4.1 Some studies where culpable and non-culpable collisions have been separated and culpability assigned by the drivers (questionnaire respondents)

Study	Culpable	Drivers/vehicles	Comments
Adams-Guppy and Guppy, 2003	59%	Truck drivers	Very diverse sample
Arthur, 1991	45.5% 44.6%	Mainly car drivers?	The subjects were instructed to report whether the police had found them culpable
Arthur and Doverspike, 1992	45.5% 42.9%	Car	—
Arthur and Doverspike, 2001	50.0%	Mainly car drivers?	—
Arthur and Graziano, 1996	62% 45%	Car	The subjects were instructed to report whether the police had found them culpable
Arthur, Strong and Williamson, 1994	44.6% 41.6% 61.0%	Mainly car drivers?	The subjects were instructed to report whether the police had found them culpable
Barkley, Guevremont, Anastopoulos, DuPaul and Shelton, 1993	50.0% (ADHD) 31.2% (control)	Mainly car drivers	—
Barkley, Murphy, DuPaul and Bush, 2002	49.3%	Car drivers	Calculated from means in table 2
Barkley, Murphy and Kwasnik, 1996	70%	Car drivers	Half the subjects had ADHD
Cellar, Nelson and Yorke, 2000	50%	Probably mainly car drivers	—
Corbett and Simon, 1992	75%	Mainly car drivers	—
Deery and Fildes, 1999	63.6%	Mainly car drivers	Novice drivers
Dobson, Brown, Ball, Powers and McFadden, 1999	52–62%, mean 55.6%	Car drivers	—
Goldstein and Mosel, 1958	50%	Probably mainly car drivers	Correlation .79 (men) and .88 (women) between all accidents and culpable

Table 4.1 *Concluded*

Study	%	Population	Notes
Harano, Peck and McBride, 1975	32.5% 36.4%	Mainly car drivers	Contrasted groups of high and low accident drivers
Hingson, Heeren, Mangione, Morelock and Mucatel, 1982	35.4%	Car, truck, motorcycle	–
Laapotti, Keskinen, Hatakka and Katila, 2001	64% (females) 60% (males)	Car drivers	–
McCartt, Shabanova and Leaf, 2003	52%	Mainly car drivers	Young drivers
Nabi, Consoli, Chastang, Chiron, Lafont and Lagarde, 2005	46%	Mainly car drivers?	Time period eight years. Only serious accidents
Parker, West, Stradling and Manstead, 1995	(52.5%) (46.4%) (56.4%)	Mainly car drivers	Active-passive classification
de Pinho et al., 2006	40%	Truck drivers	Percentage calculated from table 3, various causes to previous accidents
Roberts, Chapman and Underwood, 2004	36.9%	Company drivers	Culpability reported in five degrees. The present dichotomised value calculated from table 4
Rolls, Hall, Ingham and McDonald, 1991	56.9% (mean)	Mainly car drivers?	Values reported for six groups (age/sex). Young drivers more likely to report at fault
Sagberg, 2006	40.7%	Mainly car drivers?	–
Simon and Corbett, 1996	'half'	Mainly car drivers?	–
West and Hall, 1997	(45.5%)	Mainly car drivers?	Active-passive classification

? denotes uncertain reporting.

Note: Shown are the percentage of the total number of accident involvements that were judged as culpable and type of population studied.

Table 4.2   Some studies where culpable and non-culpable accident involvements have been separated and culpability assigned by police officers or similar officials (recorded data)

Study	Culpable	Drivers/vehicles	Comments
Barbone, McMahon, Davey, Morris, Reid, McDevitt and MacDonald, 1998	66%	Mainly car drivers	Drivers were prescription drug users, and the results indicated a higher risk for being at fault for these drivers, that is, this value is not representative for the overall population
Chandraratna, Stamatiadis and Stromberg, 2006	≈50%	Car	Partly responsible excluded. Incompatible values for data sets
Chen, Cooper and Pinili, 1995	50.7%	All types	Judgement probably from police and/or government official
Chipman, MacDonald and Mann, 2003	49.5% 41.0%	All types	At fault when police had charged the driver
Findley, Unverzagt and Suratt, 1988	58.5% (sleep apnoea) 50.0% (controls)	All types	At fault when convicted for a violation that contributed to the accident
Foley, Wallace and Eberhard, 1995	46%	Probably only car drivers	–
Gully, Whitney and Vanosdall, 1995	54.9%	Police drivers	Police panel
Kim, Li, Richardson and Nitz, 1998	56%	All types	Induced exposure method, so 50 per cent culpable assumed, but single vehicle crashes added
Leveille, Buchner, Koepsell, McCloskey, Wolf and Wagner, 1994	44%	All types	At fault when cited for a violation in crash
McKnight and Edwards, 1982	22.5% 13.0% 16.3%	Probably mainly car drivers	Accidents with convictions
Munden, 1967	39–52%	Car drivers	This study used the term 'primary vehicle', which has been interpreted as akin to culpability
Vernon, Diller, Cook, Reading, Suruda and Dean, 2002	55.8% 68.1% 68.2%	Probably mainly car drivers	Citation or noted as contributing. Calculated from data in table 5 in study
Waller, Elliott, Shope, Raghunathan and Little, 2001	25%	All types	At fault when a violation was recorded on the same day as the accident

Note: Shown are the percentage of the total of accident involvements that were judged as culpable and type of population studied.

Table 4.3   Some studies where culpable and non-culpable accidents have been separated and researchers have assigned culpability (from records or self-reports)

Study	Culpable	Drivers/vehicles	Culpability judgement	Comments
Ball, Owsley, Sloane, Roenker and Bruni, 1993	65.1%	Mainly car drivers	Three researchers	Older drivers
Brandaleone and Flamm, 1955	35–39%	Bus drivers	Probably researchers from company reports	–
Cation, Mount and Brenner, 1951	62%	Police officers	Probably researchers from police investigations	–
Clarke, Ward, Bartle and Truman, 2006	78.0%	Mainly car drivers?	Researchers from police records	–
Clarke, Ward and Jones, 1998	58.0%	Mainly car drivers?	Researchers from police records	Overtaking accidents only. Percentage calculated from table 1
Drummer et al., 2004	85%	All types	Researcher	78% for drug-free drivers, 92% for those on drugs. Cases where the driver was partially responsible added here, not used by authors
Fergusson and Horwood, 2001	(76.4%)	All types of vehicles	Researchers from self-reports	Active-passive classification
Fischer, Barkley, Smallish and Fletcher, 2007	80.3% 87.6%	Mainly car drivers?	Researchers from self-reports?	It was uncertain by whom culpability had been assigned, although it is probable that researchers did this from self-reported details
Goode et al., 1998	51.9%	Probably car drivers only	State records/researchers	Older drivers
Jick, Hunter, Dinan, Madsen and Stergachis, 1981	69.9%	Injured drivers	Researchers	20.3% of drivers were judged indeterminate regarding fault

Table 4.3   Concluded

Study	Culpable	Drivers/vehicles	Culpability judgement	Comments
Lenguerrand, Martin, Moskal, Gadegbeku and Laumon, 2008	69.2%	Mainly car drivers	Researchers from police records	Responsibility analysis was used
McGwin, Sims, Pulley and Roseman, 2000	56%	Probably car drivers only	Researchers	–
Mounce and Pendleton, 1992	84.4%	Car drivers	Accident reports by police, culpability assigned by a panel of three researchers	More than half of accidents were singles. Fatal only
Owsley, Stalvey, Wells and Sloane, 1999	60.5%	Mainly car drivers?	Researchers	Older drivers
Soderstrom, Dischinger, Ho and Soderstrom, 1993	≈66%	Motorcyclists	Researchers	–
West, 1997	(56%)	Mainly car drivers?	Researchers	Active-passive classification
Williams and Shabanova, 2003	72.2% (all) 50.0% (two-vehicle)	Car drivers	Researchers?	Fatal accidents only. 61% single vehicle accidents
af Wåhlberg and Dorn, 2007	72.4%	Bus drivers	Researcher	–

? denotes uncertain reporting.

Note: Shown are the percentage of the total number of accident involvements that were judged as culpable, type of population studied and how culpability was assigned.

Table 4.4 Some studies where company recorded culpable and non-culpable accidents have been separated, and culpability assigned by a company employee or other professional

Study	Culpable	Drivers/vehicles	Culpability judgement	Comments
Cooper, Tallman, Tuokko and Beattie, 1993	91.7% (dementia drivers) 66.7% (control)	All types	Insurance company	Culpability was assigned in percentages, the numbers presented are those who were alotted more than 50% of the blame
Lowenstein and Koziol-McLain, 2001	50.8%	All types	Traffic Accident Reconstructionist	41.8% were judged fully responsible, and those were compared with the not responsible (49.2%)
McBain, 1970	57.4% 54.7%	Truck drivers, long haul	Company	–
Moffie and Alexander, 1953	51.2%	Truck drivers	Company supervisors	–
Quimby, Maycock, Carter, Dixon and Wall, 1986	49.5% 60.0%	Motorcyclists Car	Accident investigation team	–
Rajalin, 1994	61.5%	Car drivers	Investigation team	Fatal crashes only. The non-culpable drivers had almost 50% more traffic offences on record than controls, while culpable had even more
Wilson, Meckle, Wiggins and Cooper, 2006	48.7%	All types	Insurance company	Young drivers. This value was provided by personal contact. Mean for 2005-2007, all collisions
Sagberg, 2006	49.1%?	All types?	Insurance company	Uncertain reporting
af Wåhlberg and Dorn, 2007	49.6%	Bus drivers	Transportation/insurance company	–

? denotes uncertain reporting.

Note: The percentage of the total number of accident involvements that were judged as culpable, type of population studied and source of culpability judgement.

These tables have been arranged by the four main sources of judgements about culpability in traffic research. Researchers thus use their own views, those of professionals like the police and transportation and insurance companies, and those of the drivers themselves. Here, we can see the first systematic effect, in that the highest culpability levels were given by researchers (mean 67, std 14 per cent, N=18), with companies in second place (mean 58, std 12 per cent, N=12), self-reports third (mean 50, std 11 per cent, N=33), and police officers yielding the lowest percentage (mean 46, std 17 per cent, N=18). It would seem to be the case that what degree of culpability you get is heavily dependent upon the source of this information, that is these categories of people tend to use different criteria of culpability.

It could of course be the case that some of the differences are due to real differences between populations. However, even if this was the case, and could be shown, it would not disprove the competing hypothesis about differences in criteria.

Drivers can be seen to be very probable to acquit themselves of blame. This would seem to indicate that such reporting is indeed unreliable, although that of police officers is even more so. One result that implied this conclusion was that of Hours et al. (2008), who compared the self-reported non-culpability of drivers with the assignment undertaken by researchers from files, using the responsibility analysis algorithm (Robertson and Drummer, 1994). More than thirty per cent of these self-stated innocent victims were found culpable when this method was used.

Furthermore, there is reason to believe that the self-reported under-assignment of culpability is not a general, random effect over all drivers, but associated with certain other variables. Thus, Foreman, Ellis and Beavan (1983) found that among injured people, those who reported details about their mishaps that indicated that they had caused them themselves tended to be internalizers (using Rotter's (1966) Locus of Control scale), while '... externalizers produce stories that depict themselves as blameless victims of outside factors.' (p. 224). Also, Lajunen, Corry, Summala and Hartley (1997) found that their scale for socially desirable responding had larger effects for the culpable collisions. Generally, it can be suspected that (drivers in) culpable collisions show the biases discussed for self-reported accidents in general in Chapter 2, only somewhat stronger.

Unfortunately, the use of culpability judgements even by other parties than the drivers also carries with it a risk of artefactual results, where a bias in the judgement influences the outcomes in such a way as to confirm the bias that caused the difference in the first place.[3] For example, it is commonly known amongst traffic safety professionals that young drivers are over-involved in crashes, and this knowledge can easily influence the judgement of a police officer writing a report about a collision, making it more probable that the blame will be assigned

---

3   This idea was hinted at by Janke (1991), who noted the risk of biases of police officers influencing the coding. The systematic component of this bias was not acknowledged, however.

to the younger person involved. Especially in induced exposure methods, this will cause a difference in the results that is not real. This effect can probably be found in all culpability judgements, irrespective of source, but no evidence has been found regarding this possible bias.

The general problem concerning culpability is that there is no accepted criterion that can be applied to any crash. Single crashes and shunting would seem to be accepted as this driver's fault, but thereafter, the agreement seems to stop. It can also be questioned whether it is possible, or even desirable, to create a coding scheme with so many detailed instructions as to be applicable to any crash (although responsibility analysis has taken that general path, possibly with some success). But if we cannot counter the subjective element of culpability coding by specific instructions, is there no way of getting past this obstacle?

Actually, this is possible, if, instead of focusing on the separate crashes, we take an interest in the sample. For research purposes, it is not really very important if each single case is correctly categorized, provided the majority is. For single cases, there is simply no possible way of ascertaining culpability with absolute certainty. Instead, what is needed are testable predictions about what could be expected from driver samples where culpability has been correctly assigned (to a majority of the cases), and from those where it has not. Given such predictions, it can be empirically tested whether an accident culpability criterion is correct, or whether it needs to be changed, to put more cases into one of the categories.

Such predictions have been formed, tested and confirmed (af Wåhlberg and Dorn, 2007). Here, these hypotheses about culpability criteria will be repeated and somewhat elaborated upon. The starting point is within the induced exposure method (Carr, 1969; Haight, 1973; Davis and Gao, 1995), because in this methodology, an assumption was made, which lead to testable predictions.

A Method for Determining the Correctness of a Culpability Criterion

> Apparently, as is desirable and intended, the two categories of (culpable and non-culpable) accidents are statistically unrelated. McBain, 1970, p. 516.

The basic assumption in induced exposure and responsibility analysis is that non-culpable drivers are a random sample of the driving population (Hours et al., 2008). If this assumption is turned into a definition, while removing the assumption about only one driver being to blame in a crash, it follows that a sample of non-culpable drivers should be similar to the driving population from which it was drawn. Given that the criterion for culpability used is correct, various variables will therefore have the following features:

1. Drivers' culpable and non-culpable accidents are not correlated, when exposure is held constant (because the first variable is systematically caused, while the other is random).

2. The correlation between non-culpable collisions and exposure is higher than the culpable/exposure correlation (because exposure is actually the only cause of non-culpable accidents, while culpable ones are strongly determined by the individual).

3. Drivers with non-culpable collisions only are very similar (in terms of mean and standard deviation of the group) on all variables (notably previous accidents) to those with no collisions during the same time period, while different from drivers with culpable accidents on variables which are related to collisions (because whether you have a non-culpable incident or not is random).

4. Culpable accidents correlate with the same collision predictors as 'All accidents', when exposure is held constant, but more strongly (because it is actually the culpable collisions that carry the effect for all, while non-culpable are only related to exposure).

5. Non-culpable accidents are not possible to predict with any variable when exposure has been controlled for.

6. The mean rate of responsible collisions for a group should decrease slightly over time, while non-responsible should remain the same (both held constant for exposure and changes in overall risk in the driving environment).

Most of these predictions were tested in af Wåhlberg and Dorn (2007), using two different samples of bus drivers. In one, the level of culpable accidents was about 50 per cent, in the other, more than seventy per cent. The analyses showed that the higher level of culpable crashes did have the features predicted above, while the lower level indicated that many culpable incidents had been erroneously classified as non-culpable (see Table 4.7 for correlations between these variables).

Given these features, samples with different levels of culpable accidents (due to different criteria) can be investigated, and deviations can be interpreted as being due to too lenient a criterion (which seems to be the rule so far in traffic research whenever any categorization is used). Note, however, that it is in principle possible to use these randomness criteria for determining whether the culpability criterion is actually too harsh. If non-culpable accidents are categorized as culpable, it will have the effect of obliterating the differences between the culpable sample and the population. This effect, however, is probably much harder to discern.

Furthermore, one empirical problem of these criteria that was not discussed in af Wåhlberg and Dorn (2007) is that of the time period used (see Chapters 3 and 6). It should be evident that the power to discern these features of the variables increase with lengthening of the time period used for calculations. However, a perfect categorization in terms of drivers (as belonging to a culpable or non-culpable group) can never be achieved, because a new incident can always change the status of a driver from no accidents or no culpable accidents to culpable. Therefore, for short time periods, the group of drivers with no or non-culpable accidents will always contain a few cases that have not yet realized their culpable potential. A short time period (or rather, a low mean) will also insert a small amount of artificial correlation between culpable and non-culpable crashes. Therefore, it can be recommended that samples with a mean number of accidents above two should preferably be used within this type of research.

Unfortunately, almost no research seems to have been undertaken on the correctness of culpability criteria, possibly because no methods have been available before the publication of af Wåhlberg and Dorn (2007). However, the same kind of reasoning can be found in McBain (1970), and Lenguerrand, Martin, Moskal, Gadegbeku and Laumon (2008), although the explicit testing is lacking.

As described, in af Wåhlberg and Dorn (2007), seventy per cent culpable accidents yielded the features of correct classification (non-culpable are random) described above, while 50 per cent showed various signs of having culpable and non-culpable incidents mixed. Using this as a rough guide, the research literature was searched for relevant data, with the aim of undertaking meta-analyses that could test the predictions from the randomness assumption.

Predictions, Results and Calculations Involving Culpability

> Liable collisions are a more direct indicator of risky or error-prone driving than total collisions, as the latter include those that the driver was unable to avoid.
> Wilson, Meckle, Wiggins and Cooper, 2006, p. 327.

Predictions

Given the definitions and results described in the previous section, it is possible to analyse again the research literature on culpability, but this time with certain expectations regarding the outcomes. First, it was predicted that studies that have used culpable accidents as dependent variable in parallel with all or non-culpable crashes would tend to have stronger effects for the culpable variable when the percentage of crashes judged at fault was higher, that is the mean percentage of culpable accidents would be lower in studies where this variable yielded weaker associations. This effect should be strongest for the comparison of culpable to non-culpable.

Second, it was expected that a group of correctly classified non-culpable drivers (20–30 per cent of all collisions) would tend to be similar to a group of non-accident-involved drivers, while culpable drivers would be different, on variables that have been found to relate to accident involvement.

Third, it was predicted that there would be a negative association between the percentage culpable and the correlation between culpable collisions and non-culpable crashes over studies.

Given these predictions, the research literature was searched and the studies reporting relevant results summarized.

Culpable Versus All and Non-Culpable Collisions

First, in Tables 4.5 and 4.6 can be seen a rough and ready type of quantification (dichotomization of data) of the effect of assigning culpability to different degrees in different samples. For each study, the percentage of culpable accidents was noted, and coupled with the difference in effect for the predictor variables against 'All accidents' (Table 4.5) and non-culpable accidents (Table 4.6). This difference has been denoted with a plus or a minus, showing whether culpable accidents only yielded the strongest effect or not. The expectation was for a positive association between these variables.

It should be noted that the dichotomization was difficult to undertake, as most studies used multiple predictors, and differences could be found between them. In such cases, the majority of variables were used as guide, or the overall amount of explained variance, if this was reported. In summary, both tables show that studies where culpable accidents have a stronger effect tend to have a higher percentage of collisions judged at fault, five per cent when compared to 'All accidents', and more than ten against non-culpable. The latter effect was also expected, as the non-culpable group is supposed to carry the error variance and the culpable group the real effect.

The case of McKnight and Edwards (1982) must be mentioned, though. In that study, 'All accidents' were used as criterion in parallel with accidents with convictions, which were only about 20 per cent of all, in three samples. Despite the very low percentage of culpable crashes, this variable had the stronger effect in two of the samples. The reason for not including this study in Table 4.5 was that it was not a correlational study, but a group comparison. A treatment and a control group were compared for differences on these two variables, with effects as stated. However, this situation is different from a correlational study, where one dependent mainly contains 'true cases' and the other is more or less made up of 'error' cases. In McKnight and Edwards, the comparison was between two variables with presumably equal amounts of error.

Similarity of Driver Groups

Turning to the similarity of non-culpable to accident free drivers, once again it was pertinent to undertake calculations at a group level of data reported (an instance of the secondary correlation method). Two studies were found that had included the type of data needed.

McGwin, Sims, Pulley and Roseman (2000) divided drivers into groups of at fault (56 per cent), not at fault, and no accidents, for other reasons than those stated here. Their results were as could be expected, if the criterion was too lenient; drivers in the 'Not-at-fault' group were very similar to 'at-fault' drivers on the very important variable 'Prior crash involvement', but dissimilar to those who did not

Table 4.5   Some studies that have used culpable and all accidents as criterions in parallel

Study	Per cent culpable	Effects for culpable only versus all accidents
Arthur and Doverspike, 1992	45.5%	+
	42.9%	-
Arthur and Doverspike, 2001	50.0%	-
Arthur and Graziano, 1996	62%	+
	45%	+
Arthur, Strong and Williamson, 1994	44.6%	-
	41.6%	-
	61.0%	-
Barkley, Murphy, DuPaul and Bush, 2002	49.3%	-
Barkley, Guevremont, Anastopoulos, DuPaul and Shelton, 1993	45.7%	+
Dobson, Brown, Ball, Powers and McFadden, 1999	55.6%	+
Goldstein and Mosel, 1958	50%	-
Gully, Whitney and Vanosdall, 1995	54.9%	+
Lajunen, Corry, Summala and Hartley, 1997	Not reported	+
Leveille, Buchner, Koepsell, McCloskey, Wolf and Wagner, 1994	44%	-
McBain, 1970	56%	-
Parker, 1953	Estimated at 50/50	+
Parker, West, Stradling and Manstead, 1995	(49.4%)	- (active-passive categorization)
	(56.4%)	+ (active-passive categorization)
Rajalin, 1994	61.5%	+
Vernon, Diller, Cook, Reading, Suruda and Dean, 2002	55.8%	+
	68.1%	+
	68.2%	+
Wilson, Meckle, Wiggins and Cooper, 2006	48.7%[*]	+
West and Hall, 1997	(45.5%)	+ (active-passive categorization)
af Wåhlberg, 2000	71.0%	+
af Wåhlberg, 2008a	≈76%	+

[*] This value was not reported in the paper, but was kindly provided by Dr Wilson upon request. Mean for British Columbia for 2005–2007.

Note: Shown is the percentage of culpable accidents and whether this criterion yielded stronger effects (+) for the predictors than all accidents.

Table 4.6   Some studies that have used culpable and non-culpable accidents as criterions in parallel

Study	Per cent culpable	Effects for culpable only versus non-culpable accidents
Arthur and Doverspike, 1992	45.5%	+
	42.9%	-
Arthur and Doverspike, 2001	50%	-
Arthur and Graziano, 1996	62%	+
	45%	+
Arthur, Strong and Williamson, 1994	44.6%	-
	41.6%	-
	61.0%	+
Cellar, Nelson and Yorke, 2000	50%	+
Fergusson and Horwood, 2001	(76.4%)	+ (active-passive categorization)
Parker, West, Stradling and Manstead, 1995	(49.4%)	- (active-passive categorization)
	(56.4%)	+ (active-passive categorization)
af Wåhlberg, 2008a	≈76%	+

Note: Shown is the percentage of culpable accidents and whether this criterion yielded stronger effects (+) for the predictors than non-culpable accidents.

have a crash in the study period. This indicates that the criterion used was indeed too lenient, and that denoting 56 per cent of drivers at fault is too low.

In the same study, odds ratios[4] were calculated for drivers being at fault versus not at fault and not having had an accident, when various medical problems and medications were present. If the categorization of non-culpable was correct, these drivers should be very similar to the non-accident drivers, and therefore yield similar odds ratios for the predictor variables when the comparison was made against the at fault drivers. Therefore, not only should the sizes of the odds ratios correlate between these comparison groups, but their means should also be similar.

In fact, the odds ratios correlated .76 for various types of diseases (table 2 in study, N=15), and .33 (table 3, N=18) for medications (all ratios had been adjusted for age, gender, race and annual mileage). The difference in associations for these calculations was in agreement with the means of ORs. For diseases, the difference

---

4   Odds ratios are calculated using logistic regression, and is a measure of relative risk between groups. This means that there is always a control group that has the designated base level risk odds of one, to which the other group is compared. In the example given, it was not the crash risk that was compared (which is usually the case). Instead, this was defined from the beginning, and the relative risks calculated were instead of using medications etc in these groups.

was very small (<1 per cent), while for medications, the no-crash comparison had more than ten per cent higher ORs. For the medical conditions, the result therefore indicated a fairly good categorization, while medications results did not. Why this difference occurred, when calculations were made on the same groups of drivers, has not been possible to ascertain.

It would seem to be fairly clear that ORs for medications were very dissimilar between comparison groups, while diseases were harder to evaluate, the problem being that there is no standard with which to compare to. Is .76 a fair agreement? In reality, the way to test whether the criterion is correct is to change it and do these kinds of correlations again. From the present perspective, it would be expected that the ORs of this study would become more similar if the culpability criterion was somewhat stricter.

Similar results as those of McGwin et al. were found in Rajalin (1994), where 61.5 per cent drivers were deemed culpable, but the non-culpable group still had 1.46 more violations on record than a random control group.

Furthermore, Chipman, MacDonald and Mann (2003) reported crash rates per 100 driver years for various subgroups (substance abusers and controls), in two ways; 'All accidents' and 'at fault' accidents. From these data, not at fault means of each group can be calculated, where after these can be correlated with at fault means. Given that exposure was somewhat controlled, and assuming that the criterion used for determining fault was correct, it could be expected that the means of groups for the two crash categories would not correlate. If the criterion was too lenient, on the other hand, it could be expected that the categories would have a positive correlation, because of a causal overlap. When the computation was undertaken, it was found that for the first set of data (from table 2 in study), the correlation was -.19 (N=8), and for the second set .62 (N=8). These results were thus also inconclusive, and no explanation for this strong inconsistency could be found. The difference between the sets of data was time and the enrolment in drug abuse treatment programs. However, how this could influence the results could not be known.

In the same vein, Lowenstein and Koziol-McLain (2001) found no differences between their fully culpable cases (41.8 per cent) and controls (not responsible drivers) on variables like age and sex. This is somewhat suspicious, and would seem to indicate an erroneous criterion that did not expose the weak effect that was probably present.

Correlation Between Culpable and Non-Culpable Collisions

Turning to the third prediction, that the correlation between drivers' culpable and non-culpable accidents would be negatively related to the percentage of culpable incidents in the sample across studies, a few studies reporting such values were found. In Table 4.7 the percentage of culpable accidents in each study is tabulated against the correlation between culpable and non-culpable accidents.

When the data in Table 4.7 is to be interpreted, it should be remembered that percentage culpable accidents is calculated for accident involvements (not individuals), while correlations are between the number of culpable and non-culpable crashes of each driver (both of which can be zero). Therefore, in most cases, the values for per cent and r were computed using slightly different groups of drivers.

The values presented in Table 4.7 had an association that was close to zero, and prediction number three was therefore not supported. Some further observations about this type of result can be forwarded, though.

Chandraratna, Stamatiadis and Stromberg (2006), used a crash sample, that is, only drivers with accidents (in one part of the time period investigated). This method will probably yield a stronger correlation between culpable and non-culpable than if drivers with no incidents had been included. The extremely high R squared reported (0.49) may therefore not be as deviant from the other values as might be at first thought.

However, there is also something very peculiar about this study, because the raw crash data reported for drivers did not corroborate the correlation value given. Instead, if the tabulated values in tables 3a and 3b are turned into variables and correlated, the result is $r = -.32$ (and given the massive N, highly significant) between at fault and not at fault lifetime accidents. It might be suspected that the R value was actually negative, but given the reported R squared of 0.49, R should have been .7. It has not been possible to find out what caused this result.

Table 4.7   Some studies that have correlated culpable and non-culpable accidents

Study	Culpable	Correlation
Arthur, 1991	45.5% 44.6%	-.07 -.06
Arthur and Doverspike, 2001	50%	.12
Arthur and Graziano, 1996	62% 45%	.12 .16
Cellar, Nelson and Yorke, 2000	50%	.22
Chandraratna, Stamatiadis and Stromberg, 2006	≈50%	(.49, -.7, -.32)[*]
Gully, Whitney and Vanosdall, 1995	54.9%	.12
McBain, 1970	57.4% 54.7%	-.14 .16
Moffie and Alexander, 1953	51.2%	-.12
af Wåhlberg and Dorn, 2007	49.6% 72.4%	.338 .102

[*] This association is very uncertain, as discussed in the text.

Note: Shown are the percentages of culpable accidents and the correlations between culpable and non-culpable.

For the general results in Table 4.7, however, the Chandraratna et al. value is of no importance, because none of the suggested possibilities make any difference in the interpretation, which is that there is no pattern in these data.

Arthur and Graziano (1996) reported a correlation of .24 between not-at-fault accidents and moving violations tickets. Although this could be an artefact of exposure (especially given the variable time frame used, see Chapter 6), it is nevertheless an indication that the 45 per cent culpable[5] accidents was too low a number, because at-fault accidents correlated .32 with violations, that is the culpable and non-culpable groups were not very different.

Finally, it can be noted that West's (1997) active-passive categorization seems to lead to strong correlations between these variables, as well as effects on both the predictor variables (for example, Horwood and Fergusson, 2000). It is interesting to note that the passive category contained only 24 per cent of the collisions in that study; if a similar percentage had been allotted to a non-culpable category, these events would probably not have been associated with the predictor variables. However, as the accidents were self-reported, this might not actually be true, because common method variance effects might be at play with non-culpable and/ or passive collisions too.

Discussion

> The ultimate measure of risky driving is involvement in collisions, particularly where the driver has been deemed at fault. Jonah, 1997, p. 658.

The present chapter has investigated the use of culpability for crashes in traffic safety research, found indications of a wide array of differing criteria (although these were seldom reported), suggested a method for testing the criterion of culpability, and used this on the available literature. Some of the predictions generated from a criterion of randomness of the non-culpable group would seem to be supported. In summary, the present meta-analyses showed that:

1. There is substantial disagreement between studies regarding the right percentage of culpable crash involvements, probably mostly due to differing criteria.
2. Self- and police-reported culpability is probably not dependable, as these sources yield lower percentages of culpability than the judgements from other sources.
3. Studies that have not found stronger results for the culpable group of accident (as compared to all and non-culpable) involvements tended to

---

5   How culpability was actually determined was not certain; the data was self-reported, but respondents were asked about the opinion of the police who had reported the crash.

have lower percentages of culpable crashes.

4. The results regarding similarities between non-culpable and control groups of drivers were inconclusive, although some scattered results that were not possible to meta-analyse seemed to support the prediction.

5. The prediction regarding correlations between culpable and non-culpable accident involvements was not supported.

One interesting feature of the studies referenced in this chapter is that the type of discussion given here is totally lacking in the literature, even where culpability is used as a factor in some way. A few researchers mention the problem of ascertaining culpability, but most only state from what source culpability was taken, stating no criterion, method or tests of validity and/or reliability. Mostly, there is no reporting what so ever, apart from stating that 'at fault' crashes were used (for example, Legree, Heffner, Psotka, Martin and Medsker, 2003). Among the exceptions are McGwin, Sims, Pulley and Roseman (2000), who did discuss the possibility of erroneous categorization of culpability for crashes.

Given the fair prevalence of studies that have in some way used culpability as a factor, it is therefore important that this is further investigated. Given the available evidence today, it can be suspected that the results of many studies are unreliable, but that this can be better evaluated when more evidence concerning culpability criteria and effects has been published. The present chapter should therefore not be seen as evidence for or against the correctness of the randomness criterion suggested, but mainly as a description of methods applicable to the problem of assigning culpability.

Chapter 5
The Accident-Exposure Association

It has been suggested in the literature that the crash rate per mile decreases as the number of miles driven increases... Massie, Campbell and Williams, 1995, p. 85.

Introduction

Frequent drivers are also more experienced than infrequent drivers. Parmentier, Chastang, Nabi, Chiron, Lafont and Lagarde, 2005, p. 1122.

Exposure to risk of a traffic accident is a concept that is on a par in importance with individual differences when it comes to traffic safety. Despite this, there is no consensus about how to operationalize this variable. The two basic contenders are time and distance, but the first can be conceptualized in a variety of more or less refined ways, as will be further discussed in Chapter 6. Some authors have also suggested that exposure in traffic safety should be defined as the number of risky events (conflicts or similar), that is the riskiness of the road should be taken into account for each individual. Although having some appeal, this operationalization is not really feasible, given the methods of measurement available today. It is also logically difficult to see how a variable like this could work from an individual differences perspective, as it would tend to hold constant what is supposed to be measured by individual differences.

In principle, the two concepts of exposure and individual differences have been treated by researchers as unrelated. Although people do drive vastly different mileages, and this probably to a fair degree is due to personal dispositions, this does not seem to have been investigated as a factor in collision prediction. Instead, exposure is often treated as a factor that should be held constant to better bring out the individual differences in accident proneness under similar circumstances. Therefore, a linear correction procedure is often carried out to adjust for these differences, usually using miles driven as correction.

Within traffic safety research, a few fundamental 'facts' exist that seem to have been accepted by the majority of researchers as existing and proven, despite the peculiar effects that logically follow from these phenomena. The low stability of the collision variable is one of these, where researchers have totally ignored the fact that an unreliable variable with low variance cannot be predicted. Similarly, it has been accepted that the association between accident record and exposure (usually in terms of miles driven) is not a linear function, but curvilinear, levelling off at high amounts of exposure (for example, Maycock, 1985; Daigneault, Joly and Frigon, 2002; Parmentier, Chastang, Nabi, Chiron, Lafont and Lagarde, 2005). Here, the strange discrepancy would seem to be that if this association is in fact curvilinear (exposure increases faster than collisions), why is a linear correction applied in research, often in the form of crashes per mile, that is, a proportional change? Even stranger, however, are the mechanisms behind a curvilinear crash-exposure association.

The present chapter will scrutinize the consequences of accepting this proposed curvilinearity in terms of how drivers must behave and react to create an association like that, suggest some other explanations to this phenomenon, and review the literature, trying to ascertain what explanations are actually supported in the available data.

We start by studying the claims of a group of researchers who have accepted the non-proportional accident-exposure relation to such a degree as to see it as some sort of law of nature, and thereafter have used it as an explanation for another effect, the increased collision risk per mile of older drivers. This study is instructive regarding the features of the accident-exposure association, because the shortcomings of the proposed bias for older drivers highlight what is wrong with the curvilinearity concept as such.

Accidents, Exposure and 'Low Mileage Bias'

> Examples of all these exposure data collection methods have been found ...
> but very little research has been carried out concerning the relative reliability,
> validity and costs... Wolfe, 1982, p. 340.

The supposed non-proportional association between traffic collisions and exposure (actually mileage) has been accepted as something of a law of nature by some researchers, and thereafter used as an argument that older drivers are not worse drivers than younger ones (for example, Hakamies-Blomqvist, Wiklund and Henriksson, 2005; Alvarez and Fierro, 2008).[1] Their elevated risk per mile is said to be due to their low exposure, that is, it is argued that it is natural to have many crashes per mile when driving so little, and that this phenomenon is the same for all age groups. Actually, the claim is that older drivers have more accidents per mile because they drive less than before, and probably that they would have the same rate as others if they drove more. Here, this claim will be challenged as illogical, while what little evidence there is for its basis (the curvilinearity of accidents-exposure) is explainable by other mechanisms, mainly various methodological biases.

First, it may be noted that the researchers using 'low mileage bias' (LMB) as an argument for accepting older drivers' high collision rate as inevitable are

---

1   The reason for why this claim is forwarded is probably that these researchers want to change the way older drivers are treated, that is they want more lenient treatment regarding licensing for this group.

remarkably silent regarding where (in which papers) this effect has been reported,[2] instead mainly referring to Janke (1991). However, Janke did not start from a point where she determined that the accident/exposure association was curvilinear, but rather with the inferences made from collisions per mile from differently exposed groups, arguing against this as a good measure of risk, and apparently instead proposing the total number of crashes as a more relevant estimate of the 'threat to society' of individual drivers and/or groups. Therefore, Janke too gave very few references regarding where the curvilinear association might be found.

Furthermore, when discussing the Janke paper, the LMB proponents only refer to the proposed bias of low miles suggested by Janke. However, this was but one of many possibilities discussed (Janke suggested that there may be any number of accident/exposure associations, depending upon the specific population under study), and, most importantly, Janke did not suggest that LMB was some sort of law of nature, but acknowledged that it '... seems not unreasonable to hypothesize that drivers with a low level of competence tend to have low mileages.' (p. 185), that is the opposite of what LMB believers claim. This and other possible behavioural mechanisms explaining a curvilinear association were also discussed by Maycock, Lockwood and Lester (1991) and Maycock and Lockwood (1993), including the possibility that high-mileage drivers adjusted their acceptable risk to compensate for their high exposure.

It can also be noted that the LMB activists are equally silent about research supporting their claim that high-mileage drivers tend to drive on safer roads and other qualitative differences in exposure (for example, Langford, Methorst and Hakamies-Blomqvist, 2006). Ferdun, Peck and Coppin (1967) did not find any effects for (self-reported) driving at night or miles on freeways and expressways, neither for recorded collisions nor violations, a result that was replicated for (self-reported) accidents by Maycock, Lockwood and Lester (1991). If the claim of unequal quality of exposure is made, it should at least be bolstered by some positive findings in support.

The research undertaken by the LMB aficionados in support of their basic proposition (that older drivers do not really have an elevated collision risk) is also rather peculiar. The accident reports of old drivers are compared with those of younger ones, in mileage groups, meaning that mileage is controlled for. When no remarkable differences are found, this is for some reason interpreted as evidence in favour of the proposition. First, it should be noted that the relevance of this methodology is not really explained in any of the papers using it. However, comparing the collisions of differently aged low-mileage drivers is really only a way of re-stating what was already known; drivers driving few miles tend to have more accidents per mile, without determining the causal mechanism. This research is simply beside the point. What should have been investigated instead

---

2   That is Hakamies-Blomqvist, 1998; Hakamies-Blomqvist, Raitanen and O'Neill, 2002; Hakamies-Blomqvist, Wiklund and Henriksson, 2005; Langford, Methorst and Hakamies-Blomqvist, 2006; Langford and Koppel, 2006; Alvarez and Fierro, 2008.

was whether there is a larger percentage of low-mileage drivers in the older population. If so (and with cohort effects controlled), it should point to the known fact that older drivers change their driving, driving less as they get older (Bygren, 1974; Vance, Roenker, Cissell, Edwards, Wadley and Ball, 2006), especially in situations perceived to be difficult and/or dangerous (Ball, Owsley, Stalvey, Roenker, Sloane and Graves, 1998). What is important is that more drivers belong to the low mileage population when they grow old.

The LMB argument also suffers from a basic logical and methodological flaw in the original (and actually all) data concerning the relation of exposure to collisions in terms of individual differences. When the more or less curvilinear association is formed, this is based upon data from different drivers driving different distances. However, the argument proposed by LMB enthusiasts presupposes that any driver who drives but few miles will suffer this effect. But no such effect has actually ever been shown, because the drivers driving different distances are different persons. Only if an experiment was undertaken in which drivers were randomly allocated to high and low mileages, and the curvilinearity was reproduced in their resulting collisions and exposure, this association could be used as an argument of the type forwarded by the LMB campaigners. This logical error can also be framed as the assumption that the mileages drivers drive are determined by forces that have nothing to do with their collisions. As no research on this problem seems to have been undertaken, this assumption would certainly need to be corroborated before being used.

Finally, the basic proposition of the LMB lobby ('older drivers do not have an elevated accident risk') would need some scrutiny. What is actually meant by this statement? According to their own research, it can only be interpreted as a comparison between low-mileage drivers, that is, the statement should be 'older drivers driving few miles do not have higher collision risk than other low-mileage drivers'. However, low-mileage drivers do, according to LMB advocates, have high risk per mile, because that is the reason of their research. We therefore return to the acceptance of the curvilinearity of exposure and accidents as a law of nature. Apparently, what the LMB group is saying is that if we accept the driving and collision rate of other low mileage drivers, we need to accept the older drivers' with low mileage too.

In conclusion, the LMB adherents have forwarded an idea that is not supported by what few references they usually give. Instead, they argue in a fashion that is based upon a widespread acceptance of certain 'facts' within traffic safety research. However, even if these 'facts' were actually supported by the evidence, the logic behind LMB would still make the assumption rather strange. It would seem that the proponents want to acquit older drivers of their high rate of collisions, based upon a comparison with other bad drivers, saying that they are not worse than any others. This may be so, but in that case we could not really try to limit the driving of any low-mileage group. Alcoholics, for example, would be a group in question, as they probably drive few miles per year. As the LMB lobby obviously claims that old drivers have a high rate of accidents because they drive little, the outcome is equally obviously under volitional control, rather much like the alcoholic's drinking and driving. Any counter-argument regarding the old drivers' lack of control over their situation may equally be applied to the drinking drivers, and we therefore get a very strange situation, where it is proposed that certain groups get a carte blanche for driving the way they do, whatever the consequences, because they have low mileages.

Given the doubtful referencing regarding the curvilinearity of collisions versus exposure, the LMB notion must also be evaluated against what research there is. However, before that, we need to scrutinize the concept of a curvilinear association between crashes and exposure, and consider whether such results can be explained in any other way than by some sort of (current) skill factor.

Alternative Explanations to the Non-Proportional Accident-Exposure Association

> ... it is essential to ensure that the new findings are not an aberration due to undetected methodological flaws... Langford, Methorst and Hakamies-Blomqvist, 2006, p. 575.

As described in the second section of this chapter, Janke (1991) and other writers have discussed various possible mechanisms that could create curvilinearity between collisions and exposure. However, none of them seem to have noticed that the notion of curvilinearity has rather strange consequences for how drivers need to behave to create this association. Neither did they discuss the possibility that this effect is at least in part due to methodological problems. Here, the logic behind non-proportionality will be described, as well as a number of methodological biases that can be suspected to exist in the available data.

When researchers have studied the closely related concepts of exposure and experience, the first has usually been conceptualized in terms of mileage, and the other in years of driving. However, they have seldom been analysed as complementary variables, and, possibly therefore, a strange contradiction between the findings for these parameters seems to have gone unnoticed. A curvilinear association between yearly mileage and collisions would seem to indicate that, if you drive little, increasing your driving with a few thousand miles per year would make you a remarkably better driver. However, when experience is tabulated against accidents, it has repeatedly been shown that it makes little contribution beyond the first years (Mayhew, Simpson and Pak, 2003; Dorn and af Wåhlberg, 2008), or even months (McCartt, Shabanova and Leaf, 2003), because the crash rate is strongly reduced, while at the same time, mileage per month increases for young drivers (McCartt, Shabanova and Leaf, 2003).

As a result of the supposed non-proportionality of the collision-exposure association, a rather peculiar mechanism must exist; the usually over-learned and automatic skill of driving must deteriorate rather fast if the amount of driving is reduced. A 'low mileage effect' can hardly be interpreted in any other way, as the initial increase in skill (decrease in risk) is apparent during the first few years or even months, as noted. A similar, but non-linear, mechanism could be that a driver must be driving a certain number of miles per year to even reach a high level of proficiency, and that those doing lower mileages therefore never reach this far. The latter mechanism is even less likely than the first, as it also does not accord with the initial strong increase in skill when driving is first undertaken. Neither of these possible mechanisms has been investigated.

Would it be possible that a skill that is often described as 'automated' (after a few years driving) is actually so superficially learned that it is markedly reduced if not constantly updated by driving thousands of miles each year? It should also be remembered that for old drivers, most of them have hundreds of thousands of miles of experience. No evidence has been found regarding such a phenomenon, and it has certainly not been presented by the LMB proponents, although it is a logical pre-requisite for their claim.

The curvilinearity of the accident-exposure association can, on the other hand, to some degree be questioned simply as an artifact of methodology. Apart from the suggested difference in quality of exposure (high-mileage drivers drive on safer roads), three other mechanisms can be forwarded as explanations of the reported curvilinearity; the use of self-reported data, confounding by experience, and co-variation with accident proneness.

For the exposure side of the association, it has been shown by Staplin, Gish and Joyce (2008) that self-reports of mileage are biased in such a way as to, to some degree, explain the curvilinearity. Apparently, this is due to a tendency of over-estimating high mileages, and under-estimating low ones, although this effect is still somewhat uncertain, as discussed in Chapter 2 . For collisions, it has been shown in Chapter 2 that the opposite effect is at work; drivers with many accidents on record tend to report fewer ones, probably due to memory difficulties and/or social desirability. This mechanism would therefore interact with the exposure report, creating (part of) the curvilinearity that has been found, when data is self-reported. It can be noted that the researchers arguing for a lack of over-involvement of older drivers in collisions have all used self-reported data (for example, Hakamies-Blomqvist, Raitanen and O'Neill, 2002; Langford, Methorst and Hakamies-Blomqvist, 2006; Alvarez and Fierro, 2008) in their tests of LMB.

Given that self-reported data (for mileage and collisions) creates a stronger curvilinearity than what actually exists, several predictions follow. It can be expected that the strongest non-proportionality should be found in reports with self-reported data for both accidents and exposure, because both these variables contain an artefactual non-linearity. Thereafter, studies using only one self-reported variable should have an in-between non-proportionality, while those with objectively measured data only should have almost proportional results. The reasons for thinking that the last type of results will not be totally linear are the two other possible explanations for non-proportionality, which are real effects that create a certain amount of curvilinearity in the data.

The collision-exposure association could also be confounded by a correlation between age/experience and exposure. As driving tend to increase with age for car drivers (but decrease for motorcycle riders; Taylor and Lockwood, 1990) those with little experience (and thus a high risk) would mainly be found in the low mileage categories. This was probably so for the study by Ferdun, Peck and Coppin (1967), shown in Figure 5.1, where the oldest drivers on average drove twice as many miles as the youngest ones.

Finally, it can be suggested that there is indeed a causal relation between exposure and accidents, but that the effect is in the opposite direction from that assumed by LMB devotees. What is proposed is simply that bad drivers tend to drive fewer miles, as suggested by Janke (1991), which could be thought to be some sort of compensatory mechanism. This hypothesis has apparently never been tested.

Here, it can be remembered that older drivers often report that they reduce their driving because of the difficulties they experience (for example, Ball, Owsley, Stalvey, Roenker, Sloane and Graves, 1998; Vance, Roenker, Cissell, Edwards, Wadley and Ball, 2006). In other words, the reduction in mileage is, at least partly, due to a perceived decline in safety. For some reason, the LMB disciples have disregarded these statements from the older drivers and proposed the opposite. Actually, the original proponent would seem to have paid no attention to her own research in this matter (for example, Rimmö and Hakamies-Blomqvist, 2002). But it might also, of course, be the case that the LMB fans think that the various

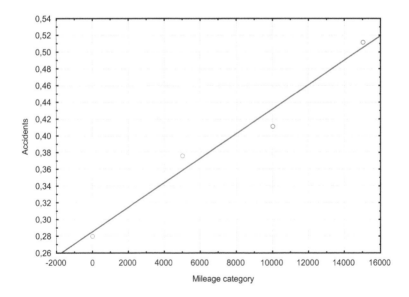

Figure 5.1 **The association between annual mileage category and mean number of accidents reported in Lefeve, Billion and Cross (1956), table 7. Recorded accidents and self-reported mileage**

bodily maladies of older drivers only cause a reduced amount of driving, which in its turn causes the increased collision rate. Such a mechanism would in principle be possible, and represents a testable prediction, but no one seems to have realized this and tested it.

In summary, there are several possible mechanisms that could explain parts of the curvilinearity believed to exist between accidents and mileage, probably leaving very little of this effect if they were all controlled for. We therefore turn to the available evidence regarding the collision-exposure association, keeping in mind the proposed methodological biases suggested.

Studies on the Accident-Exposure Association

> ... mileage data are typically self-reported and therefore subject to considerable error. Janke, 1991, p. 184.

As noted, the curvilinear association between collisions and exposure is well accepted within traffic safety research, apparently to such a degree that the LMB zealots, whose argument rests upon the assumption of this being true, find it superfluous to give any references to where this effect has been found. For the present purpose, database searches, snowballing and the other techniques of finding relevant research were employed. The aim was to find association measures and/ or data on exposure versus collisions where the degree of curvilinearity could be studied. As usual, for the aims of this book, most of the relevant values were found by reading papers on individual differences in traffic safety in general, as an exposure variable is sometimes included without this being the main aim of the study, and this term therefore not being given in the title and/or key words.

First, various association figures for accidents and exposure (that is mileage) were gathered and are displayed in Table 5.1. It can be seen that most values are fairly low. This, however, might be an effect of curvilinearity, and these values can therefore only be taken as very rough indications of the true effect. In fact, they only indicate how much of the variance can be explained by a linear component. More important, controlling for mileage in studies on individual differences in collision liability does not seem to yield positive effects in terms of explained variance, as can be seen in Table 6.2.

To study the evidence for non-proportional increase, data were taken from studies containing values for groups with different amounts of exposure. These data were all in the form of mileage classes and/or means of groups, with the mean number of accidents given for each such grouping.

In most studies, the data were presented in terms of mileage categories (for example, 5,000–9,999 miles). This created a problem which number should be entered as the mileage for this group; the number required was really the mean of this category of drivers. To further complicate matters, the highest category was often only denoted as a certain number and all higher values. Therefore, it was

Table 5.1 **Studies on the association between accidents and exposure; the data types and sources, and the effects found**

Study	Accident source	Exposure measure	Statistic	Effect	Comments
McBride, 1970	State record?	Mileage in one year	Pearson?	-.233	High violation/ accident drivers
McGuire, 1972	Self-report	Mileage	Pearson?	.18 .23	–
Quimby, Maycock, Carter, Dixon and Wall, 1986	Self-report	Miles 1 year Miles 3 years Miles 5 years	Kendall correlation	.14 .13 .12	Data from table 4. Accidents for five years
Stutts, Stewart and Martell, 1998	State record	Mileage	Odds ratio	1.08	–
West, 1995	Self-report	Mileage six months Mileage past year	Spearman	.17 .05	Young drivers in first sample
West, Elander and French, 1993	Self-report	Annual mileage	Pearson	.12	–

? denotes uncertain reporting.

Note: Associations calculated on the level of the individual driver. All exposure self-reported.

decided to use zero as a starting point for these studies, because using the midpoint of each class would leave the highest category without any such point. This method made the values chosen somewhat imprecise as estimates of the actual mean, but was used for all figures with the description 'Mileage category', while studies reporting the mean for a group have a caption reading 'Mean mileage'.

The associations between crash means and mileage were thereafter computed for each study, with results that can be seen in Table 5.2. Also, the data was graphically rendered in Figures 5.1–5.7. It can be seen that no study reported an increase in collisions that was proportional to the increase in mileage. The correlations between the means of accidents and mileage were all very high. However, these associations were artificially inflated due to the aggregation of data, as the size of the correlation was very much dependent upon number of categories used (N in Table 5.2).

What relations could be expected in these data, given the suggested methodological biases? If the data had been for the individual driver, the correlations for the recorded accidents would have tended to be stronger, due to the linear component, but given the aggregations used, this hypothesis could not be tested. Also, a stronger linearity would probably yield lower constants in the regression equations. This expectation was not borne out, as the data sets using recorded accident data instead had the highest constants (see Table 5.2).

Table 5.2   The association between collisions and mileage in the studies presented in Figures 5.1–5.7, in the same order

Study	Accident source	N	Correlation	Linear increase	Type of driver/vehicle
Lefeve, Billion and Cross, 1956	State record	4	.99	Accidents = 0,2852 + 14,6400 × Mm	Car drivers
Ferdun, Coppin and Peck, 1967	State record	15	.65	Accidents = 0,1282 + 4,3331 × Mm	Car drivers
Liddell, 1982	Self-report	5	.97	Accidents = 0,0459 + 7,5588 × Mm	Car drivers
Taylor and Lockwood, 1990	Self-report	7	.96	Accidents = 0,1101 + 41,8100 × Mm	Motor cyclists
Maycock, Lockwood and Lester, 1991*	Self-report	6	.98 .95 .86 .70	Accidents, males = 0,1061 + 3,7511 × Mm Injury accidents, males = 0,0114 + 0,4019 × Mm Accidents, females = 0,0927 + 3,5344 × Mm Injury accidents, females = 0,012 + 0,28512 × Mm	Car drivers
Lourens, Vissers and Jessurun, 1999	Self-report	5	1.00 .98	Accidents, drivers with fines = 0,0886 + 2,8732 × Mm Accidents, drivers without fines = 0,0519 + 2,2605 × Mm	Car drivers

Note: All mileage data was self-reported. Unless otherwise stated under Linear increase, the accidents and drivers used were of all categories. Shown are the source of accident data, the number of group means, the correlation and regression equation (constant plus the increase per million miles) for these, and the type of drivers.

*The data for accidents was also published in Maycock and Lockwood, 1993.

However, there is one effect in these data that would seem to corroborate the self-report bias hypothesis. In Figures 5.8–5.9 can be seen a re-calculation of the associations presented in Figures 5.5–5.6, showing the number of crashes per mile in different mileage groups. The reason for showing this different type of calculation was that the visual impression of figures of crash versus mileage associations might be misleading, due to the scales used and the initial jump in accident figures at low levels of exposure, meaning that a regression line will not pass through origo. These effects mean that it will be difficult to interpret the angle of the line, as there is no standard to which it can be compared. When accidents per mile are tabulated against mileage groups, a totally linear association between crashes and miles would create a flat line, while a curvilinear association would still be curvilinear, but decreasing instead of increasing. It can therefore be seen in Figures 5.8–5.9 that the injury accident data actually had much more stronger linear associations with the mileage means than the all accidents variables, something that might not have been apparent in the first type of rendation. This result would be expected from the findings reported in Chapter 2 regarding the influence of severity of crashes on their probability of being reported.

The study by Taylor and Lockwood (1990) was not included in the comparison of equations, as it used motorcyclists as subjects, and these are known to have a substantially higher risk of crash involvement.

In studies by Massie, Campbell and Williams (1995) and Massie, Green and Campbell (1997) curvilinear associations were reported between collisions and exposure, but these results are questionable, as they were derived using different databases and forming means for age groups. Actually, it is difficult to understand how exposure was associated with crash involvement in these studies.

Parmentier, Chastang, Nabi, Chiron, Lafont and Lagarde (2005) apparently found a curvilinear association in their self-reported data, as it was adjusted by a log-linear procedure to correct for this problem. However, no values were reported on the initial relation.

Finally, Gresset and Meyer (1994b) reported an increase in number of accidents by twelve percent per 10,000 kilometres. Other calculations also indicated a non-proportionality. However, the exposure data was, as usual, self-reported.

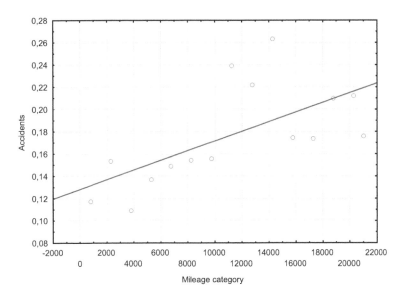

Figure 5.2 **The association between annual mileage category and mean number of accidents reported in Ferdun, Peck and Coppin (1967), table 8. Recorded accidents and self-reported mileage**

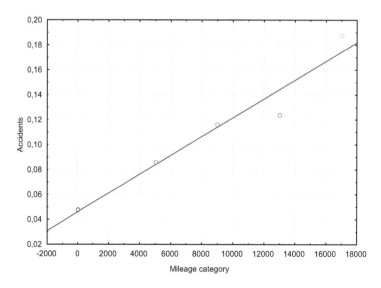

Figure 5.3   The association between annual mileage category and mean number of accidents reported in Liddell (1982), table 3. Both variables self-reported

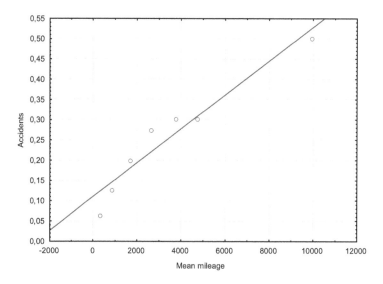

Figure 5.4   The association between annual mean mileage and mean number of (motor cycle) accidents reported in Taylor and Lockwood (1990), estimated from figure 1. Both variables self-reported

Note: Used by kind permission from TRL.

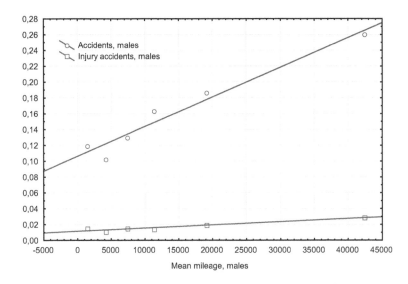

Figure 5.5   The association between annual mean mileage and mean number
of accidents for male drivers reported in Maycock, Lockwood
and Lester (1991), table 3. Both variables self-reported

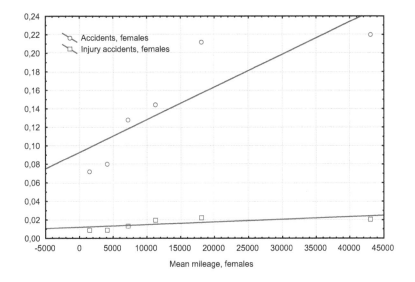

Figure 5.6   The association between annual mean mileage and mean number
of accidents for female drivers reported in Maycock, Lockwood
and Lester (1991), table 3. Both variables self-reported

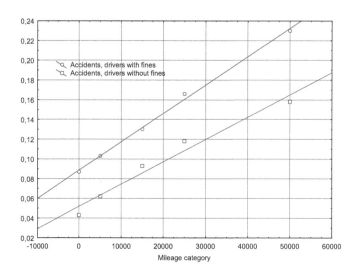

Figure 5.7   The association between annual mileage category and mean number of accidents for two categories of drivers (with and without traffic fines) reported in Lourens, Vissers and Jessurun (1999), table 3. Both variables self-reported

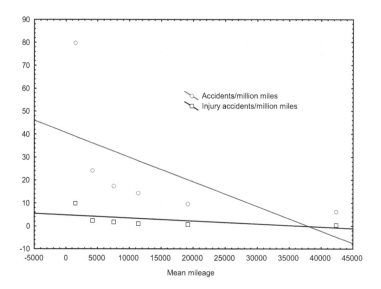

Figure 5.8   The association between annual mean mileage and mean number of accidents per million miles for male drivers reported in Maycock, Lockwood and Lester (1991), table 3. Both variables self-reported

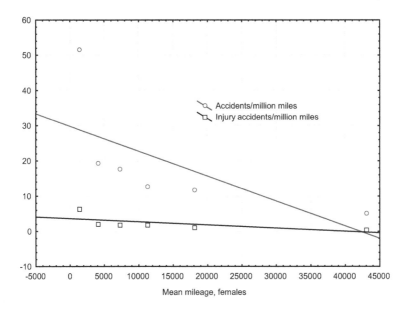

Figure 5.9   The association between annual mean mileage and mean number of accidents per million miles for female drivers reported in Maycock, Lockwood and Lester (1991), table 3. Both variables self-reported

Discussion

> ... their accident frequencies hardly increase with mileage at all ... No satisfactory explanation has yet been given for this effect. Maycock, 1997, p. 242.

Many authors have discussed the concept of exposure within individual differences in traffic safety (for example, Chapman, 1973; Risk and Shaoul, 1982; Wolfe, 1982; Janke, 1991; Maycock, 1985) and the various problems that result from different operationalizations etc. However, judging by the dates of the studies shown in Table 5.1, interest in the exposure problems has been waning the last decade or so. As with so many other research problems discussed in this book, the traffic research community seems to have preferred to leave the exposure enigma unsolved, and gone on to greener pastures.

It can be concluded from the presently reviewed and re-analysed data on the association between accidents and exposure (in reality mileage) that:

1. None of the studies has found that collisions increase proportionally with mileage.

2. There is actually very little evidence (that is few studies reporting) that the association is curvilinear.
3. There is even less evidence that high-mileage drivers do more of their driving on low-risk roads, and/or that this influences the accident-exposure association.
4. All studies on the collision-mileage association have used self-reported mileage.
5. The results reported for injury accidents by Maycock, Lockwood and Lester (1991) are compatible with a self-report bias in accident data.
6. None of the studies reported on here has controlled for experience.

It can therefore be concluded that there is scant support for the suggested bias of using self-reported accidents in the data presented here. However, the data is sparse and difficult to evaluate, and the other types of bias suggested have not been possible to review, as they have not been tested, and their effects on the association remain unknown. Until these biases have been investigated, the question about how mileage and collisions are related remains open, because the data published so far is not trustworthy. As for the LMB argument regarding older drivers' accident rate, there is little reason to pay any attention to it, as it rests upon assumed but largely unproven or disproven assumptions about results and the methodologies used for attaining these, when they exist at all.

It is possible that self-reported mileage creates some of the reported curvilinearity between collisions and exposure, but only Staplin, Gish and Joyce (2008) have specifically addressed this problem. Replications of this result are needed.

The other proposed mechanisms that might create a real non-proportionality in the collision-exposure association need to be discussed. As currently interpreted, the reported curvilinearity means that a driver who drives a lot reaches a high level of proficiency, and needs to keep driving a lot to maintain it. Apart from this conclusion being at odds with other findings, it is challenged by the competing explanations of confounding with experience and accident proneness. It should be pointed out here that these two mechanisms are very different from the current interpretation. Both of them say that drivers with high mileages have low collision rates, but not because they are driving a lot right now, but because they have been driving much in their lives, and that drivers with a low risk will tend to drive even more. The current explanation, though, seems to say that the accident follows the amount of driving, regardless of other factors. However, this is of course an interpretation of statements by various researchers, who have not spelled this out in detail. These researchers may not agree with how their words have been rephrased into a mechanism, but if this were the case, it would be interesting to know exactly how they conceptualise the behaviour of drivers that lead to a real non-proportionality between collisions and exposure, without any of the factors proposed here.

Apart from the dearth of objective data regarding collisions and mileage, it can also be concluded that there is precious little research regarding alternative ways of conceptualising exposure, the main contender being time spent driving. We do not really know how this different denominator could influence, for example, the outcomes of research into individual differences in crash liability. This will be discussed further in Chapter 6.

The methodological consequences of the findings for the present chapter are not very revolutionary, despite that an established 'fact' of traffic safety has been challenged and to some degree repudiated. It may actually be fortuitous that few researchers have bothered with correcting for mileage in their studies on individual differences in traffic safety, as we do not really know the consequences of this method. Other types of exposure control, on the other hand are essential, for reasons that are discussed in the next chapter.

Chapter 6
Constructing a Driving Safety Criterion

Introduction

> ... we choose to utilize crash involvement as an outcome measure. The premise of this variable as an outcome measure is that individuals are more likely to be involved in a crash if they drive unsafely (sic!) ... Newnam, Griffin and Mason, 2008, p. 636.

One of the most basic problems of transportation is safety, and the task of trying to predict who is a dangerous driver has probably been one of the most common goals of all psychology-based studies within the field. However, the very important terms of 'dangerous' and 'safe' are in fact very poorly defined, researched and used within an individual differences perspective. Although an unspoken consensus seems to have been reached to accept almost any type of conceptualization in publications, this does not mean that anything goes in terms of scientific quality. What can easily be is seen is that there is much variation, and no empirical research in terms of comparisons between different conceptualizations.

Here, the term 'conceptualization' is used to denote the way researchers think about their dependent variable in terms of its various properties (like culpability, the effect of exposure, reliability etc). The lack of an explicit conceptualization leads to the problem that is the subject of this chapter; how do we construct the (individual differences in) collision record parameter? For most researchers, this does not appear to have been experienced as a problem, and it is hardly ever discussed (af Wåhlberg, 2003a). However, as should be evident from the present chapter, there is good reason to do so.

If accident proneness, or liability, or safety, were to be predicted, it would seem to be natural to want to compare how many collisions different drivers would have if they drove under the same environmental conditions, that is the same roads, vehicles, weather etc, but also the same number of miles. Under such idealized circumstances, we could truly say that the differences between drivers in numbers of crashes would be due to personal dispositions of some kind. As no such idealized driving conditions are likely to ever exist, naturally or by experiment, researchers need to take the next best solution, controlling for these influences as well as they can.

In the present chapter, a number of different ways of calculating accident record will be discussed regarding two features of importance; how 'safety' is conceptualized when a certain method is used, and what the likely consequences are of using it. As a tool for this undertaking, it is assumed that the users of these methods for some reason, stated or not, have assumed that their way of doing it has some kind of advantage when it comes to predictive power and some kind of safety application, like the selection of professional drivers, or the identification of behaviours most likely to cause crashes.

There seem to be several different problems that (different) researchers try to tackle when they construct a driving safety criterion; exposure being the most prevalent, but also collision types, the relation of accidents to other behaviours, and the mathematical relationships between all these. In general, however, these problems are not discussed, and it can therefore only be guessed why certain methods of calculation have been used.

Three broad classes of exposure approach can be identified in individual differences research; those who calculate collisions per exposure unit, those who use exposure unit per accident, and those who utilize an absolute number of crashes, without any reference to exposure (this was also discussed in Chapter 3). The problem of the collision criterion is mainly the differences between these methods, that is, how to handle (mainly amounts of) exposure, where different versions can have very different results. For example, the problem of older drivers' accidents discussed previously shows the difference between two of the possible versions; although their crashes decrease per year, they increase per mile. So, if safety is conceptualized as the number of collisions a driver causes in some sort of absolute sense within a time period, older drivers are very safe. But if it is defined as how long (in miles or hours) you can actually drive without having an accident, they are rather dangerous. So, depending on how you handle exposure, the results from your research might be very different.

However, there are also other criterion problems, chiefly that of what time period to chose for the calculation. As was seen in Chapter 3, the reliability of collision record is low when car drivers and short time periods are used, even though the stability of accident-causing behaviour would seem to be very strong. Also, the present chapter will discuss a few approaches to the collision criterion that do not fit the broad classification made above with regards to exposure. Some authors have inserted various kinds of weighting, thus making different accidents more or less important than their counterparts, while others mix their collisions with alternative safety variables, something that is treated in its pure form in Chapter 7.

But starting with the exposure problem, it will be shown that even this seemingly basic choice has been made rather differently between researchers. As very little data is available on what the effects are when choosing different time periods, the main part of this section will discuss the logic and possible consequences of such choices.

Length of Time Period

> A short observation period tends to distort and highlight individual differences
> in accident involvement. Sass and Crook, 1981, p. 178.

If a study of differential driver safety is to be undertaken, for what time period
should accidents be gathered for the subjects? As with all criterion problems
discussed here, the choice of length of time period for the calculation of collision
frequencies (or time intervals) has received little attention from the researchers
making these choices. Among the very few papers found so far that discuss this
problem in some way are McKenna (1983), Kloeden, McLean, Moore and Ponte
(1997) and af Wåhlberg (2000; 2004b). In these papers, the notion of an optimal
time period for the calculation of accident frequencies in relation to a predictor
was discussed in various ways.

Why is the time period used for gathering crash data of importance in collision
prediction studies? First, it can be noted that if no specific time period is chosen,
and no other exposure control used (hours, mileage), then we will mainly have a
measure, not of differential driver safety, but of differential driver exposure. This
problem was discussed in Chapter 3, and will be treated again in a section below,
so here it will suffice to underscore the result of using no exposure control when
calculating a safety criterion; a mixed measure of an individuals' propensity to
cause collisions and to drive.

Returning to the question of what reasons there are for choosing different time
periods, the commonly given answer would probably be that a longer period would
increase the reliability of the dependent variable, and an increase would therefore
always be a positive thing, as long as it did not bring with it some practical
drawback, like attrition of samples, or memory problems. This assumption of
increased reliability would seem to have been vindicated in Chapter 3. However,
it can be suspected that an increase in time period will not always yield a better
prediction, due to changes in the predictors over time.

The length of the optimal time period is therefore thought to be determined by
two opposing factors; the longer the period used, the more stable will the collision
history as a measure of the construct 'accident liability' be. Simultaneously, the
longer a time period used, the less predictive power the predictors will have, as
all human variables change,[1] and a measure taken at a certain time will decrease
in validity with time. The two opposing principles of data aggregation (Rushton,
Brainerd and Pressley, 1983) versus instability of the independent variables
will thus result in a change in strength of the correlations between crashes and
predictors with the time period used as basis for the collision frequency. It will first
increase, until an optimum is reached, and then decrease.

The empirical results in this area are scarce, as researchers very seldom have
used criteria calculated over different time periods. Four different time periods

---

1　Although of course some, notably sex, change very little.

(2–5 years) were used in af Wåhlberg (2000), and six in af Wåhlberg (2004b), with no interpretable results.[2] However, in a study by Quimby and Watts (1981), seven out of ten significant predictors (driving errors, speed, visual acuity etc) had stronger associations with the two-year period than with the three-year period. One explanation for this somewhat counterintuitive result could be that the accident data were self-reported, and that a memory effect made the quality of the data so much worse for the longer period that this was not equalled by the increased variance. However, it is also possible that some of the predictor variables had low reliability, that is, were unstable over time.

Similarly, in af Wåhlberg (2006b), the only time period to yield consistently significant correlations using collisions was two years, despite tests with from one to nine years. This effect, however, was later shown to be due to the low reliability of the predictor;[3] when it was aggregated, predictive power was evident for more than five years, although it was strongest for the same period when celeration data had been gathered (almost three years, af Wåhlberg, 2007d).

The concept of the optimal prediction time period being short for variables with low reliability and high for others can also find some backing in a study by Ball, Owsley, Sloane, Roenker and Bruni (1993). This study found that the longest period tested (five years) yielded the best results. As their predictors were mainly vision variables, this does make sense, as changes in eyesight are usually fairly slow processes, and these variables therefore probably had high reliability.

It can also be pointed out that the time period used should be related to the predictor variables in terms of not preceding them. Some predictors (not many, but the principle is important anyway) have specific onsets, instead of being more or less permanent from before licensing and onwards. This was pointed out by Hu, Trumble, Foley, Eberhard and Wallace (1998) regarding medications, but may also be true for such events as acute stress situations, like the death of a relative or friend. Here can also be noted the very interesting method used by Haselkorn, Mueller and Rivara (1998), who studied the effects of brain injury on driving. To separate the effects of the injury from other differences, these authors held several variables constant for cases and controls, among them previous accident and violation records. This ingenious method is probably only applicable to the study of conditions with specific onsets, but might be very important for these, as the cases of Haselkorn et al. had an elevated crash rate even before their brain injuries. Such results also raise interesting questions about the stability of general medical syndromes that might be of importance for traffic safety; it could be that less specific measures of health/illness would constitute better predictors of collision involvement (af Wåhlberg and Dorn, 2009b).

A method that is similar to that of Haselkorn et al. was used by Bédard, Dubois and Weaver (2007) when they controlled statistically for previous crash record

---

2   The samples were small.

3   The predictor used was celeration, which is the mean or the sum (depending upon the calculation) of all speed changes undertaken by the driver (af Wåhlberg, 2008b).

in a study of effects of cannabis on driving. The difference between this method and that of simply comparing the collision records of users and non-users is worth pointing out; the latter shows whether people who use cannabis are more dangerous drivers, but if this is the case, it still has not proven that it is because of the use of this drug (or any other) that the difference occurs. The method of Bédard et al. is much closer to proving such a thing as the detrimental impact of cannabis, and should be more widely used.

In summary, the question about what time period to use for calculating a driving safety criterion has received little attention and whether an optimal period for such constructs in relation to the predictors actually exists is not known. What is needed are some fairly large studies using accident data with good validity for a long time period (at least ten years) and several independent variables with reliability figures for similar time periods. With such data, it could be ascertained whether collision prediction can indeed be optimized from the relation between the reliabilities of dependent and independent variables.

Calendar time in relation to accidents has apparently attracted interest from researchers in traffic safety, in terms of often being used to keep exposure somewhat constant. Exposure in terms of actual driving has also received some attention, mainly in terms of mileage, as we have seen in Chapter 5. Here, mileage and some additional ways of conceptualizing and controlling for exposure will be discussed.

Accidents per Exposure Unit

This accident rate was a crude measure, devised at the time to make practical use of the data available. Simon and Corbett, 1996, p. 761.

Most people would probably agree that a safe driver has no collisions, while one with many is not safe. Thereafter, disagreement reins. The basic problem is due to exposure to risk, based on amount of driving. If you do not drive at all, like many older persons, it is fairly easy not to have accidents as a driver. But what if you drive only a little, or very much? Could a driver be considered safe if he has some collisions on record and drive 100,000 miles a year, or a 100,000 miles in a lifetime? What if exposure is calculated as hours of driving instead? And finally, which method of exposure measurement should be used in relation to the number of accidents, which one has accumulated, to use as criterion? This section will take a look at the various possibilities that exist and, when possible, present data where the differences between methods can be discerned.

The exposure problem is most often (in collision prediction studies and similar undertakings) handled in two complementary ways, by using accidents per time unit, and per mile unit. In effect, what is done is to calculate the number of collisions per year for a variable time period, or use a fixed number of years. The last is the most common approach. Next, if mileage data is available, accident rate can be calculated per this unit, or the effect held constant. Similarly, hours of driving can be used for the same refinement. Furthermore, some researchers have tried adding different qualities of exposure, like night-time driving to predict collision number, or to hold this influence constant. However, we start with the most basic type of exposure control, the fixed time period.

Accidents per Year

The most common method of holding exposure to traffic accident risk constant in individual differences research is using collisions for a specified time period as criterion (af Wåhlberg, 2003a). But as drivers differ enormously regarding their yearly mileage, this would seem to be a very crude way of controlling for exposure.

What is amazing is that it actually works, in the sense that effects are often found on variables tested as predictors of individual differences (see reviews by Signori and Bowman, 1974; Lester, 1991; Peck, 1993; Clarke and Robertson, 2005), which to a large degree use this method (af Wåhlberg, 2003a). The only possible explanation seems to be that these individual differences are much stronger determinants (explaining more variance) than the differences in risk on different roads, and the amount driven. The latter conclusion would seem to be vindicated by the association figures reported in Chapter 5 (Table 5.1), showing very little variance in accident record being explained by exposure. The differences in mean level of risk due to the use of different roads, on the other hand, would seem to be hard to quantify. In principle, it would need a driving diary and/or GPS-based study over several years to really ascertain how large this effect is. No such study has been found, only self-report-based ones.

It would seem that it has been assumed by researchers claiming the 'low mileage bias' effect as one of the explanations for old drivers' increased collision rates that there are considerable differences between drivers' driving risk exposure with regards to the roads driven. However, as with the mileage bias in itself, these researchers give precious few references for such a claim. Therefore, it can, for the time being, be questioned whether there is any significant influence of road exposure on individual accident record, as compared to the variance explained by the individual's behaviour.

As will be discussed under the heading of 'no exposure control', the strength of the individual driver's collision-causing behaviour would seem to come across even when exposure is totally disregarded (although it is hard to know what is really predicted in most studies, due to various other methodological shortcomings). Before that, however, we will have a look at what the effects are when exposure is controlled in terms of amount, at a somewhat more sophisticated level than simple calendar time. How is it even possible to predict something (individual differences in accident record) that is so hopelessly tangled with a very different variable? There are two possible explanations.

First, we assume that the effects found in some studies using collisions per year as criterion are true, that is they are not due to common method variance or other artifact mechanisms. Thereafter, it can be suspected that some variables used for accident prediction, personality for example, are not only related to the way people drive, but how much. Unfortunately, it is uncommon that correlations between mileage and predictors are reported, but some data on this has been found (see Table 6.1). Although these results do not present a very comprehensive or convincing picture, the possibility of an interaction between collision predictors and mileage remain. If this explanation is actually correct, it would mean a rather different view about what accident liability or proneness is. Research into this problem is certainly called for.

The second explanation, which can happily co-exist with the first, is the one described above, that drivers' accident proneness is so very strong that it comes through the exposure noise and creates the bulk of the variation to be explained. This second possibility would seem to be the one that most researchers in traffic safety are operating under, although it does not seem to have been stated in any of the papers reviewed here.

There is one advantage of using a fixed time period for the collision variable, which controlling only for mileage does not have; long-term changes in the driving environment are held constant. Although this is probably not a grave problem under normal circumstances, it might have some influence if there have been a sudden change, like the oil crisis of the early seventies, when driving speeds were reduced in the US to save fuel, with a measurable reduction in crashes as a result (Chu and Nunn, 1976).

Having noticed the possibility of a methodological problem (co-variance between exposure and predictors) in the use of exposure control by time period, we turn to more advanced ways of calculating an accident criterion.

Accidents per Mile

Mileage must be considered to be a much better exposure control factor than a fixed time period. Indeed, it could in some ways be thought of as fairly ideal. However, in principle, it can be criticized for differences in risk level for drivers mainly using different roads. However, as noted in Chapter 5, although this possible bias has been discussed and claimed by 'low mileage bias' researchers as an explanation for older drivers' high collision rates, it has apparently never been shown to have any measurable impact outside of self-reports.

The basic question for the present chapter is how the safety criterion influences the results of research. We need therefore to turn to the literature again and compare results where mileage has been controlled for in some way, and where it has not. When discussing this, two different research approaches can be discerned. Accidents can simply be divided by mileage and used as a criterion (for example, Maki and Linnoila, 1976), which can be called a direct control, or mileage can be

Table 6.1   The correlations between mileage and various variables that have been tested as predictors of traffic accidents

Study	Statistic	Effect	N	Predictors	Comments
Deffenbacher, Lynch, Oetting and Swaim, 2002	Correlation	.41 .21 .22 .17	290	Frequency of anger Intensity of anger Frequency of aggression Frequency of risky behavior	–
Groeger and Grande, 1996	r	.159	308	Self-assessment of driving skill	–
Guibert, Duarte-Franco, Ciampi, Potvin, Loiselle and Philibert, 1998	t	p<.01	1,784	Medical conditions (any)	Medical examination
McBride, 1970	Correlation	.050 .108 .148 -.007	75	Ascendancy Responsibility Emotional stability Sociability	Gordon Personal Profile. Extreme violation/accident sample
Sullman, Pajo and Meadows, 2003	Correlation	.002 .111 .062 .097 -.090 -.009	382	DBQ violations Speed Mild social deviance Risky driving Safety climate Experience	Truck drivers. Probably the same data as in Sullman, Meadows and Pajo (2002; 2004) DBQ= Driver Behaviour Questionnaire
West, 1995, Study 1	Spearman	.23 .04 .11 .14 .04 .08 .09	316	DBQ Alcohol consumption Alcohol/safe driving Drunk driving DWI SMQ Attitude to violations	Young drivers SMQ= Social Motivation Questionnaire

Table 6.1   Concluded

Study	Statistic	Effect	N	Predictors	Comments
West, 1995, Study 2	Spearman	.42	376	Speed	–
		.13		Alcohol consumption	
		.23		Alcohol/safe driving	
		.19		Drunk driving	
		.08		DWI	
		-.01		SMQ	
		.29		Attitude to violations	
		.20		Experience	
West, Elander and French, 1993	Pearson	.14	108	Social deviance	–
		-.06		Thoroughness	
		.05		Type-A	
		.30		Speed	
		.11		Driving deviance	
		.02		Age	
		-.43		Sex	
West, French, Kemp and Elander, 1993	Correlation	.32	48	Speed	Objectively measured
West and Hall, 1997	Correlation	.33	406	DSQ	DSQ= Driving Style Questionnaire
		.26		Attitude	
		.00		Deviance	

Note: All mileage self-reported. All predictors self-reported, apart from Guibert, Duarte-Franco, Ciampi, Potvin, Loiselle and Philibert (1998) and West, French, Kemp and Elander (1993). Shown are the statistic used, the effect of mileage upon other variables, the number of subjects and the other variables used.

entered as a co-variate in some type of multivariate statistical calculation. These methods probably yield somewhat different results, although this does not seem to have been tested, and it is beyond the aims of the present text to explore this question. Instead, we turn to research where exposure has been controlled in either of these ways, especially that where effects have been reported in parallel with and without mileage control. For the present chapter, the main interest is to compare the zero-order results with those controlled for exposure, especially as differently calculated criteria are very seldom compared within a study. The results from such studies can be seen in Table 6.2.

In general, the results would seem to indicate very small effects of controlling for mileage, as could be expected from the fairly low correlations reported between exposure and collisions. In most of the comparisons, the influence would even seem to be negative, in terms of less variance being explained. This was certainly so for Quimby, Dixon and Wall (1986), where twelve out of fifteen reported correlations for a five-year accident period, and nine out of ten for three years, were stronger without mileage control (probably self-reported data). The conclusion is thus in line with the one presented about accidents per year; the individual differences factor is so much stronger that it cuts through the exposure effect. However, as exposure in all these studies was self-reported mileage, it is possible that this is a confounding factor that obliterates the real association.

Accidents per Time Unit of Driving

The use of collisions per hour of driving is very unusual within individual differences research, but Chipman, MacGregor, Smiley and Lee-Gosselin (1993) reported a stronger association for time than distance with crashes (self-reported data), and different patterns of associations with other variables (1992).

Here, there would seem to exist an important difference between time data from private and company sources; the first is probably more different from mileage than the latter, because it is influenced by speed. Company data, on the other hand, often report the time for salaried work, which is not always the same as actually driving. Such a difference should be very apparent for taxi, bus and to some extent truck drivers. When using or interpreting time data, it is therefore important to consider how it actually differs from mileage information.

Thus, in af Wåhlberg and Dorn (2007), bus drivers' non-culpable accidents correlated .153 with hours worked, while the association for culpable incidents was slightly weaker. However, holding these influences constant had little impact on the correlations of these categories of collisions between time periods (as reported in Chapter 3). This might be an effect of a weaker association between mileage and accidents as compared to actual driving hours and collisions, where these bus drivers' hours worked were more indicative of their distance travelled than their exposure time.

Table 6.2   Studies that have reported effects for accidents per year and accidents controlled for mileage

Study	Statistic	Zero order effect	Mileage controlled effect	Difference	Accident source	Predictor	Comments
Owsley, McGwin, Phillips, McNeal and Stalvey, 2004	Risk ratio	1.08	1.40	+	State record	Effect between groups for intervention	
Parker, Reason, Manstead and Stradling, 1995	Rate ratio	0.41 (all accidents) 0.20 (active accidents) 0.19 (active shunts) 0.07 (active ROW violations)	ns 0.23 0.22 0.11	- + + +	Self-report	DMQ-thoroughness	Also controlled for experience, age, sex and DSQ. Data from table 8 and text. DMQ= Decision Making Questionnaire
Parker, Reason, Manstead and Stradling, 1995	Rate ratio	1.03 (all accidents) 1.05 (active accidents) 1.11 (active ROW violations)	ns ns ns	- - -	Self-report	DSQ-speed	Also controlled for experience, age, sex and DMQ. Data from table 8 and text. DSQ= Driving Style Questionnaire
Quimby, Maycock, Carter, Dixon and Wall, 1986	Kendall correlation	.12 .11 .11 .12 -.19 -.15 ns -.14 .12 .12 -.12	.11 ns ns ns -.20 -.14 -.12 -.12 .13 .11 -.12	- - - + - + - + - Tie	Self-report	Static visual acuity Visual field Kinestetic visual acuity Kinestetic visual acuity Glare sensitivity Glare sensitivity Stroop test Switch test Hazard perception test Hazard perception test Hazard perception test	Data from table 4

Note: All mileage data self-reported. Shown are the statistic used, the effect before mileage control, the effect with mileage controlled for, the difference between these effects, the source of crash data, and which predictor was used.

In the end, there is very little information available about the relationship between time exposed and collisions (and most of it is self-reported, anyway). Given the new in-vehicle data recording technology now available for scrutiny, this is something that will probably change in the future.

Accidents per Weighted Exposure

As noted in the collision per mile section, controlling for distance leaves the possible differences in driving on different roads uncontrolled for. In principle, the different risk levels between, for example, rural and city roads, and day versus night driving might confound the prediction of individual differences. However, the data regarding such an effect is sparse.

In one of the few exposure-quality-control studies that have used recorded data, Kraft and Forbes (1944) calculated the accident risk exposure for each of 482 street-car drivers, based upon the differences between routes (collisions/mile), between different times of day and week, and seasonal variations. The final result was a ratio between a driver's actual and expected accidents, the latter being based upon the amount of driving undertaken weighted by the five factors stated above.

This extremely fine-grained and time-consuming work did, however, not yield any impressive differences. The authors themselves reported that the collision ratio and the simple accidents per mile index correlated 'highly', without specifying this number, although they also pointed out that there were definitive differences. Unfortunately, the authors did not report on any comparative analyses for predictors of these different collision criterion variables. It should be pointed out that this method has its greatest applicability when there are large differences in the quality of exposure between drivers, like permanent night or day shifts. It can also be mentioned that Häkkinen (1958) performed risk calculations between routes for his bus and tram drivers, and concluded that the effect of unequal exposure was negligible.

On the other hand, Maycock and Lockwood (1993) reported that driving in the dark had 2.6 times as high a risk of accident as daytime driving (self-reported data). However, they also pointed out that there were several other factors beside darkness as such that could explain this association, factors they had not controlled for. Type of road did not have an effect.

It is also possible to control for differences that are known to exist at the group level. This means that, instead of weighting an individual driver's record by the type of driving undertaken by this particular person, a value is used that is a mean for all drivers known to share a particular characteristic. This was done, for example, in the studies reviewed by Janke (1994), where adjustments were made for, amongst other variables, mean income, and collision and citation level of the area where a subject was living.

Group Comparisons; Contrasts and Case-Control

Grouping Design

As in all research on humans, what population you draw your sample from, and how you select it, will have a definitive effect on the outcome of your study. In this section, subjects/samples that can be regarded as non-normal, in the sense of their means and standard deviations of collisions, are discussed. This includes the case-control design samples, which are usually not recognized as non-normal, but actually are contrasted groups.

Using drivers' accidents during a specified time period as an outcome variable does not necessarily mean that an association measure is used in the statistical analysis. Instead, drivers can be grouped and differences between these groups calculated. However, how the groups are formed can differ quite a lot, probably due to the researcher's theoretical, methodological and statistical approach. Most often, however, these are not reported. The most basic and straightforward type of grouping in traffic safety is drivers with collisions versus drivers without collisions for a specified period. This crude analysis has the advantage of being insensitive to non-linear effects, and it is probably easier to get a significant result using this method, as compared to more advanced strategies.

However, it should be remembered that having no crashes within a certain time period does not mean some sort of absolute safety, only that the risk for these drivers would seem to be lower than that of the other group. There is probably no such thing as a totally safe driver. Therefore, a grouping of accident- and non-accident loaded drivers for a specified time period is a comparison of two parts of a continuum, where the time period and overall risk in the population determines how large a percentage of the total each group will have. Also, the time period used for the crash-free drivers will have an impact on the sample. As the time period chosen grows longer, the more extremely safe the no-crash sample will be.

Contrasted Groups

In some studies, the effect of grouping is further increased by the use of extreme samples (for example, Harano, Peck and McBride, 1975), that is, the end parts of the continuum of driving risk, especially the high end. This means that the collision group is constructed from drivers with several crashes in a fairly short time. The effect of contrasted groups is thus to further increase[4] the variation in the sample, and therefore the statistical power, by including mainly cases that are very high or very low on the variable of interest, in our case traffic accidents. A similar effect can be had by using some kind of extreme population, like criminals, who have many driving incidents. The result is inflated variation and effects (for

---

4   As long as the mean number of accidents in the population and time period under study is lower than .50, the case-control method will yield an increase in variation.

example, Meadows, Stradling and Lawson, 1998). It should be pointed out this means that the predictive power you can have in such samples will not be present in normal samples, that is the results are not really applicable outside the population the sample was drawn from. This means that when using 'normal' drivers with many collisions versus those with none, the effects found are only valid for a comparison between those extremes, but not the large bulk of drivers in between the extremes.

An example of the use of contrasted groups is Violanti and Marshall (1996), where the control group was a random sample of the population that was purged from drivers with any collisions on record over the last ten years. Therefore, the comparison was not between a dangerous group (those with a recent accident) and a normal group, but a super-safe group, and the results in some respect inflated. This happens whenever subjects are not randomly drawn from the general population.

Similarly, or differently (it is hard to tell), Asbridge, Poulin and Donato (2005) used a group of young drivers without a driver's license as the baseline (control) group in a logistic regression when comparing experience as crash predictor. Here, it is hard to understand exactly what the authors wanted to show. In some respect, it would seem like a sure bet to use as baseline a group that should have no exposure and thus no collisions, but this should also mean that the odds ratio becomes infinite, if the number of crashes were compared.[5] However, this was not what happened, because 3.4 per cent of the unlicensed drivers apparently reported that they had been in an accident as a driver in the preceding year. This instead leads to other conceptual difficulties, because the meaning of such a baseline group becomes indeterminate, as apparently at least some of the unlicensed drivers had been driving. Furthermore, those driving under such circumstances may be a very different group from the general population, as some of their driving was probably illegal (although they could have been accompanied by a licensed driver).

In general, it can therefore be said that contrasted grouping designs often have problems of validity of the results, where effect sizes are often inflated, and where the selection of drivers sometimes makes the outcomes unintelligible. The issue of grouping will also be discussed under the heading of mixed criteria discussed in a later section in this chapter.

Case-Control

In contrast to the continuous association measures preferred by psychologists, medical researchers most often use a methodology where 'cases' are compared to 'controls', which at present translates into drivers with and without collisions, most often within a certain time period (for example, Gresset and Meyer, 1994).

---

5   If, on the other hand, other variables than collisions were compared between such groups, the comparison would be meaningless, because the control group would not be a safe one, as the only reason they had no crashes was because they were not allowed to drive.

This type of thinking has its roots in health-related problems, where it is often the case that you are sick, or not, that is there are two clearly defined states. This is in contrast to the accident proneness thinking, where it is often assumed that there is an underlying, continuous trait of proneness, which manifests itself more or less within an individual. Although these different methods (contrasts and associations) do not necessarily differ regarding their assumptions in this matter, their ordering of subjects by matter is clearly different.

It should be noted that when using the case-control method, it is still important to control statistically or match for exposure (as done by Guibert, Duarte-Franco, Ciampi, Potvin, Loiselle and Philibert, 1998; Lings, 2001), as this probably for some part determines which group a person belongs to (see Table 6.1). This is probably less so when associations between variables are tested within one group, although this, as with many other features of the collision calculation problem, does not seem to have been tested, that is there are no studies which compare which method is the more robust one when it comes to lack of exposure data.

The use of case-control also has interesting methodological consequences that are similar to the ones encountered for the other methods described here. In the end, what cases and controls you chose is a mirror image of how you conceptualize safety. For example, Jick, Hunter, Dinan, Madsen and Stergachis (1981) compared a group of injured drivers who were judged to be at fault for their accident with injured passengers, as well as a mixed group consisting of drivers not at fault, drivers whose fault could not be established, and people injured in collisions where it was not determined whether they had been driving. The passenger group would seem to be a fairly well-conceived control group, similar to the not-at-fault party in two-vehicle accidents, who in induced exposure techniques are assumed to be a random sample of drivers (af Wåhlberg and Dorn, 2007). However, it could be questioned whether the passengers in injury-producing vehicle collisions really are a random sample. It could be suspected that they are more similar to the drivers they were riding with than a truly random sample would be, and therefore any comparison would tend to underestimate the effect.

The second control group of Jick et al. was also interesting, as it mixed three potentially fairly distinct sub-groups, where only one (not at fault drivers) could be said to yield an understandable comparison. As the status of the other two groups was not determined, a great deal of uncertainty was inserted into the comparisons. Therefore, the effects found for these groups, as compared with individual cases, cannot really be interpreted.

Conclusions

It is difficult to reach a conclusion about what the differences in effects are between studies using association measures and those using group differences, by sheer logic. No comparative studies have been found. In principle, a comparison of the ranking of different risk factors when using these two main methodologies could answer this question, but this is in itself a major undertaking (made even more complicated by the differences in data sources and time periods often in existence between these methods) that was outside the scope of the present text. That different grouping principles may have different results can be seen in Lenguerrand, Martin, Moskal, Gadegbeku and Laumon (2007).

In summary, we therefore do not really know how grouping drivers (where 'accident-free' drivers are the criterion) influences the results, as compared with other methods, apart from inflating the results. However, the examples given here should have shown that it is fairly easy to run into conceptual difficulties in the group method tradition too, and that this is often associated with exposure, just as the association method is. Also, what conclusions can be drawn are not only dependent upon the effect size, but also very much upon what kind of groups have been used. Such considerations are often lacking in group studies.

The practical effect of an inflated result is that when it is applied to a normal population, it has less power than thought, if the original effect is taken at face value. For example, if a contrasted grouping design was used, and it is found that the predictor could categorize 70 per cent of the drivers correctly in accident and no-accident groups (50 per cent would be random), we would not have this level of correctness if we tried it on a normal population, but maybe 55–60 per cent, less than half. The real effect would still be the same, but when differences become more fine-grained, it is more difficult to make the right categorization.

Exposure Units per Accident

Time Intervals Between Accidents

The inventors and main proponents of the 'time per accident' calculation approach are Shaw and Sichel (1961; Shaw, 1965; Sichel, 1965). Their arguments for this method and against collisions per time unit are many and often rather hard to follow. From the beginning, it can be noted that their approach has not caught on[6] (for an exception, see Williford and Murdock, 1978), for reasons that can be guessed at, although of course not determined.

The basic principle of Sichel (who seem to have been the one responsible for the statistics of the Shaw and Sichel research) was that collision rate should be calculated as a mean of the time intervals between successive incidents, a value known as t-bar. Apart from this, Shaw and Sichel also used two methodological principles, which they claimed to have determined empirically for optimal results:

---

    6   There are some researchers who have used a similar method (for example, Foley, Wallace and Eberhard, 1995), but it is uncertain whether Shaw and Sichel have inspired them. Some statisticians have also used this type of construction, but not for predicting individual differences (for example, Mintz, 1954b).

1. All incidents were included regardless of blame and severity.
2. There was a learning period (of individual length) when a driver started within the bus company where the research was undertaken, during which t-bar was not stable.

It can be noted that the first principle is mainly a question of a higher mean in the sample, something that, as shown in Chapter 3, has a positive effects on the intercorrelations between time periods, that is, reliability. The second is more unusual (actually, no one else seem to have used it) and interesting. As no exact cut-off score would seem to have been stated,[7] a certain degree of subjectivity seems to have been introduced into the methodology. It may even be speculated that this second corollary method was the real reason for the t-bar approach. Without the t-bar, the initial period of learning could not have been eliminated, as it could not have been calculated.

One practical problem, which ensues from the t-bar approach, is that a safe driver cannot be measured with any reasonable level of confidence. For an accident per time-methodologist, a driver with no collisions is a safe driver, while for Shaw and Sichel, he is indeterminate.[8] In the driver populations those authors worked with, there were very high rates of accidents, but the method still created a sizeable problem; in Sichel (1965) more than 20 per cent of the drivers in the sample used had to be excluded due to 'excessive standard errors', that is very few collisions with differing time periods between them. For car driver populations in the industrialized world of today, the method would be useless, as their crash rate is very low as compared to that of the drivers Shaw and Sichel worked with. Furthermore, excluding drivers from a prediction study is always a very suspect method.

It can also be noted that although many practical and statistical arguments were given for the superiority of the t-bar approach, no empirical test seems to have been undertaken whether this method was actually better for predicting collision involvement as compared to the standard collision per exposure method. What was reported on, however, was the split-half correlation of t-bar over time. This

---

7   In contrast to the lengthy statistical reasoning, the actual data was presented in rather vague terms. For example, the question of whether severity of accidents should be included in some way in the calculations was, in their first paper, described in three sentences, of which one was the conclusion; '... this approach did not yield any very helpful results.' (Shaw and Sichel, 1961, p. 5).

8   Davey (1956) probably used a similar method, dividing number of years of driving with number of crashes. However, the following quote, although difficult to understand, would seem to indicate that some other logic was also applied; 'This method is not entirely satisfactory for it puts into the same group the careful driver who will only have a single accident in 40 years, but has it in his first year, and the more reckless man who has been driving for ten years and has had an accident in each of them.' (p. 71).

reached .62 when the variables had been log transformed (.49 raw). Although no mean number of accidents was given for this sample, it can be estimated to be about eleven when the 20 per cent safest drivers had been excluded. A comparison with the results in Chapter 3 would seem to imply that these correlations were actually somewhat lower than what would have been expected from the meta-analysis of split-half correlations. Therefore, the conclusion by Sichel (1965), that '... a high split-half reliability coefficient of r = 0.62 was obtained ... thus justifying the use of average time interval t as a criterion ...' (p. 15) does not seem to have been justified. Actually, the conclusion would not have been sustained by any value short of 1, because what Sichel was arguing was that his method was superior to 'the old approach', and this can only be proven by getting results from both methods from the same data and comparing them.

In conclusion, it would seem that the t-bar method in itself does not seem to have been of any scientific or practical importance, neither actually nor potentially. The concept of an unstable learning period in the beginning of driving could, however, have some merit, although it presently would seem to lack empirical as well as theoretical support.

Turning to other research in this vein, Caird and Kline (2004) used a somewhat similar approach when taking the number of collision-free kilometres at the time of the study and using it as an outcome measure. This method was also used by Groeger and Grande (1996). As number of accidents was used simultaneously as dependent variable in these studies, it is possible to compare the associations found, which can be seen in Table 6.3. No definitive effects (differences or similarities) can be ascertained in this material, and more data is therefore needed, especially as some of the associations are in the same direction, which they should not be.

Another example of a kind of time interval criterion, apart from the ones described in the opening of this section, is Carty, Stough and Gillespie (1998).

Table 6.3   Some studies that have used number of accidents and number of accident-free kilometers as dependent variables in parallel

Study	Statistic	Accident effect	Kilometers effect	Accident/ kilometers source	Predictor variables
Caird and Kline, 2004	Correlation	.02	-.07	Company record	Errors
		-.19	.15		Planning
		.13	-.06		Environment
		-.16	-.08		adaptions
					Speed
Groeger and Grande, 1996	r	-.079	.185	Self-report	Self-assessment of skill
		-.095	-.127		Assessment of novice skill

Note: Shown are the type of statistic used, the effect of the predictor variables on number of accidents and accident-free kilometers and what predictor variables were used. All predictors were self-reported.

In this most peculiar study, as many as eleven criteria were used in parallel (and twenty-one predictors). One of these dependents had some resemblance to the collision-free kilometres variable; time since last crash (presumably self-reported). It was unfortunately not stated how the criteria correlated between themselves, but secondary correlations could be calculated, and are shown in Table 6.4. It can be seen that, in general, the results were very heterogeneous, inter-correlating rather poorly. For the 'time since accident' variable, it would not seem to be more or less so than any other, the only remarkable result being that its effects correlated most strongly with those of the change in driving style post collision variable. As none of the variables was defined in any way, it is unfortunately difficult to really understand these results, and they should rather be seen as yet another example of how different criteria yield different results, which in its turn point to the importance of rigorously defining safety. It might also be noted that some researchers have studied the time intervals between accidents in their own right, but most often with the aim of ascertaining whether a crash has any influence on the probability of having another (for example, van Nooten, Blom, Pokorny and van Leeuwen, 1991; Blasco, Prieto and Cornejo, 2003).

So far, little evidence or reason for using the interval method instead of per exposure unit has been published. However, there is one possibility; Maisto, Carter Sobell, Zelhart, Connors and Cooper (1979) calculated the time between convictions for driving under the influence for alcoholics, and found that there was a decrease, that is the time between them grew less with time. Here, there seems to be a possible benefit of the time interval between accidents approach; trends over time can be studied. However, if there are trends, and if so, what they mean is yet to be determined. Yet another similar case has been found, but this will be discussed under its own heading of odd methods below, as it cannot really be said to be an interval method. Instead, we turn to those studies that pay no heed what so ever to exposure.

No Exposure Control

Although amount of exposure (mileage, hours) has consistently been found to be (weakly) associated with number of collisions at the individual level, it is actually rather common to find studies which use all the accidents of a driver as the criterion, without any reference to exposure at all; a figure of more than 20 per cent (of more than a hundred studies) was reported in the review af Wåhlberg (2003a), and many more have been found since then.[9]

---

9   Stewart, 1957; Goldstein and Mosel, 1958; Hormia, 1961; Sobel and Underhill, 1976; Kishore and Jha, 1978; Gonzalez-Rothi, Foresman and Block, 1988; Gulian, Glendon, Matthews, Davies and Debney, 1988; Aldrich, 1989; Matthews, Dorn and Glendon, 1991; Barkley, Guevremont, Anastopoulos, DuPaul and Shelton, 1993; Bener, Murdoch, Achan, Karama and Sztriha, 1996; Barkley, Murphy and Kwasnik, 1996; Wu and Yan-Go, 1996; Carty, Stough and Gillespie, 1998; Blanchard, Barton and Malta, 2000; Deffenbacher, Huff,

Table 6.4 The secondary correlations between the correlations between predictors and different criteria in Carty, Stough and Gillespie (1998)

Variable	Time since crash	Accident seriousness	Responsibility	Preventability	Change in driving style	Speed convictions	DUI	Dangerous driving convictions	Cost of accidents	Number of claims
Number of accidents	.69	-.76	.39	.22	.71	.56	.40	.58	.61	.39
Time since crash		-.49	.50	.40	.73	.37	.16	.54	.23	.28
Accident seriousness			-.24	-.26	-.31	-.36	-.19	-.55	-.61	-.29
Responsibility				.43	.52	.38	.34	.28	-.02	-.01
Preventability					.41	.49	.03	.25	-.18	.25
Change in driving style						.42	.36	.52	.26	.27
Speed convictions							.40	.45	.24	.50
DUI								.73	.53	-.03
Dangerous driving convictions									.61	.09
Cost of accidents										.09

Note: Criterion variables were number of accidents, time since last crash, seriousness of crashes, responsibility of the driver, the preventability of the accident, change in driving style after accident, number of speed convictions, convictions for speed, driving under the influence and dangerous driving, and the cost of the accidents. The time period used for these variables was possibly three years, or variable. The dependent variables were not defined beyond their names. Independent variables were medical, stress, personality and cognitive variables. All data probably self-reported. N of secondary correlation=21. Original N=58. Subjects were a mix of various professional drivers.

As a slight variation on the theme of no exposure control, the study by Beck, Yan and Wang (2007) might be mentioned, where all crashes of the drivers, in ages from 16 to 65+, were used as dependent variable, but controlled for frequency of driving. Yet another, similar, method is to use different time periods for different groups, for example, an accident in recent months, as compared to drivers without collisions for several years (McGuire, 1956a; 1956b; Plummer and Das, 1973). This is similar to some grouping strategies described below. Similarly, a set period of time can be used for one group and a variable one for another, as practiced by Friedland et al. (1988). However, the mean numbers of years of driving were very similar for these groups, so this peculiarity might not have been very important.

Yet another method is the misleading, stated use of a set time period, which cannot have been used for all subjects. This was reported by Cellar, Nelson and Yorke (2000). Participants had been asked to report collisions for a ten-year period, but the mean age was only 20.9 years, making it impossible for all to have driven vehicles for this period, thus yielding a variable time period (the variable was given as 'Total' accidents). The only other explanation would be that the respondents reported about crashes they had experienced as passengers.

Given the large prevalence of this phenomenon, it is important to point out the ensuing problem, which can be summarized as; what is predicted? A criterion variable consisting of all the collisions that have ever happened to a driver, regardless of whether he/she has been driving for two or twenty years, would seem to be very different from the number of crashes in the last three years, for example. The thinking behind this type of criterion is hard to pinpoint (as no-one who has used it has discussed it), but would seem to implicate that exposure has no significance at all. Such a position becomes almost fatalistic; a driver has a certain number of accidents whether he drives a lot or not and this number cannot be influenced. However, this logic runs into immediate trouble, because that means that the younger drivers will need to have had all their allotted incidents when they are studied, or else the calculations will not be appropriate. The use of all collisions for an individual driver as a criterion variable therefore seem to be illogical, and also very vulnerable to memory and socially desirable responding effects (as pointed out for ADHD studies in Chapter 2), as this kind of calculation (or lack of it) most often is used in conjunction with self-reports.

Lynch, Oetting and Salvatore, 2000; Häkkänen and Summala, 2000; Iversen and Rundmo, 2002; Barkley, Murphy, DuPaul and Bush, 2002; Adams-Guppy and Guppy, 2003; Deffenbacher, Lynch, Filetti, Dahlen and Oetting, 2003; Wallace and Vodanovich, 2003; Dahlen and Ragan, 2004; Dahlen, Martin, Ragan and Kuhlman, 2005; Knouse, Bagwell, Barkley and Murphy, 2005; de Assis Viegas and de Oliviera, 2006; McEvoy, Stevenson and Woodward, 2006; de Pinho et al., 2006; Gnardellis, Tzamalouka, Papadakaki and Chliaoutakis, 2008. Some of these cases are uncertain, as the reporting was not always conclusive regarding how the criterion had been constructed. The total number might for example, have been converted to accidents per year in some cases, or held constant for number of trips etc, although this was not reported.

As example of what the consequences of the use of a variable time period might be, is given in a study by Barkley, Fischer, Edelbrock and Smallish (1990). Here, the researchers compared the accidents of an ADHD group with controls, for their entire time as licensed drivers. The age range was eight years for the combined sample, and only 13 per cent were licensed at the time of the study. It could be thought that if the time period for collision summation were from being licensed for both groups, they would be comparable. However, the control group had a mean age that was a year lower than the ADHD group, and it can therefore be suspected that this difference was also present to some degree for those with licenses. If that was so, the ADHD group would have a fair percentage more exposure, as the mean time for being licensed can only have been a few years given the low age of the subjects. The (non-significantly) higher percentage of drivers with accidents, as reported by their mothers, in the ADHD group could therefore be suspected to be due to more exposure, although the details needed to ascertain this were not reported.

Variable time periods for criterion formation should for these reasons not be used in most cases. There are two possible exceptions to this rule; first, if the predictor is cumulative over time and measured for the same time period as collisions (lifetime) for each driver, like driving violations could be (but not sex, for example). However, associations between such variables will probably be over-estimated, due to artificially inflated variance (which in its turn is dependent upon differences in exposure).

Second, if the comparison is made between groups matched or similar for age or years of driving experience (for example, Weiss, Hechtman, Perlman, Hopkins and Wener, 1979; Shope, Waller, Raghunathan and Patil, 2001; Vassallo et al., 2007[10]), or the time period for accidents (for example, Lings, 2001; Nasvadi and Vavrik, 2007), or controlled for amount of exposure (Parmentier, Chastang, Nabi, Chiron, Lafont and Lagarde, 2005; Zhao, Mann, Chipman, Adlaf, Stoduto and Smart, 2006), then the results might be comprehensible and valid. For example, a study by Ysander (1966) matched cases' and controls' time of holding a licence, exposure period and (self-reported) amount of driving. Such a meticulously constructed comparison is unusual today.

In conclusion, the variable/lifetime strategy for constructing the collision variable would seem to be impossible to use without getting results for associations that are misleading. When comparing differences between matched groups, the method seems to be viable (for example, Soderstrom, Dischinger, Ho and Soderstrom, 1993).

---

10   This study used a cohort with very small differences in age, but the time for licensure (equaling the accident time period) still had a standard deviation that was more than one third of the mean, so control for exposure by time period was in place but poor. On the other hand, some sort of further exposure variable was used although it was not perfectly clear how this was constructed.

Why do researchers use lifetime accidents as an outcome measure? As this choice never seem to have been discussed, it is hard to know. The only apparent advantage is the higher variability and statistical power. What this power is used to detect is another matter. It is probably not individual differences in accident proneness under equal circumstances, but rather differences in liability to exposure. However, it can be suspected that the use of variable time periods is simply an oversight in most cases; the researchers have not paid any attention to this part of their research.

Different Accident Types, Weights and Mixed Criteria

> Eleven accident variables were collected for analysis. Carty, Stough and Gillespie, 1998, p. 6.

Leaving the subject of exposure control behind, we turn to the issue of collision types, the mixture of accidents and other variables, and the mathematical methods sometimes used, where different categories are assigned different mathematical weights. From these various parts, a multitude of criteria can be constructed.

Accident Types

> Number of serious crashes was a count that included each individual subject's crashes that were alcohol-related, at-fault, or single-vehicle. Shope, Waller, Raghunathan and Patil, 2001, p. 651.

As discussed in Chapter 1, taxonomies of collisions can be created, with the aim of using only certain categories of crashes as criteria of safety in relation to the specific predictor used. The available research using different categorizations were largely covered in that chapter, and only a few cases of special interest will be described here.

A study by McBain (1970) is very interesting from a criterion-constructing perspective, as it seems to be the only instance of a peculiar technique of mixing different time periods, but also used as many as fifteen different criterion variables. Unfortunately, the whole undertaking was very poorly described, and the construction of the accident criteria only hinted at.

In general, McBain used an aggregation method that is akin to factor analysis, adding the standardized scores of highly correlated measures. However, this was applied to the (originally 29) criterion variables as well as the predictors. This principle resulted in the amalgamation of the (standardized) collision variables crashes in the last five years and crashes since employment (subjects were truck drivers). Given the results in Chapter 3, it is easy to understand that these two variables would correlate very highly, even if they had not shared variance due to time overlap. But, as with all the other criteria described here, it can be asked how this researcher defined safety when this criterion was constructed?

Severity of accident was used as categorizer (no collision, minor, major) by Matthews, Dorn and Glendon (1991) and Matthews, Desmond, Joyner, Carcary and Gilliland (1996). In both reports, the result was that minor accidents yielded differences, while major ones primarily did not, probably due to the higher variance of the small incidents variable.

In general, the use of specific collision types as outcomes in studies is fairly uncommon. As with all other criterion choices, these are seldom discussed and comparative research largely lacking, apart from that reported about culpability in Chapter 4 (which was probably not intended as such anyway).

Weights

> Each recent accident was given a number of 'responsibility points' ranging from 0 ... to 4 ... and the points summed for each driver. Simon and Corbett, 1996, p. 776.

In the majority of studies on differential collision involvement, the accident involvement of each driver was used as the basic unit of the criterion, usually, but not always, divided by some type of exposure. However, a few researchers have developed criterion models where collisions are further manipulated mathematically, in effect giving different weights to different crashes. The weights used include culpability, severity and accident recency (in relation to the study undertaken). The difference between such an approach and categorizing lies in the latter including and excluding collisions as events from the calculation, while the former changes the value of one for each incident included to some percentage of this. For example, if we had two drivers with one collision each, where one was culpable, and the other not, in a categorization by culpability design, the first would get a value of one, the other zero. In a weighting by culpability design, the culpable driver could have any value above zero, while the other would be somewhat lower, say 0.8 and 0.4, depending on what mathematical weights had been chosen.

A weighted criterion construction approach was, for example, used by Moffie, Symmes and Milton (1952) in a study of truck drivers. First, the total number of accidents for a driver's career seems to have been used, although this was not explicitly stated, that is a variable time period. Second, they calculated the time between collisions (months/accident). However, it is the third feature that really set them apart and makes this study interesting. In the truck company records collisions were assigned culpability (preventable/non-preventable), and Moffie et al. agreed that this was an important distinction. However, they also thought this dichotomization was a bit crude: 'In almost every non-preventable accident examined, there were factors operating or conditions present which *lessened* or prevented the possibility of a driver foreseeing and avoiding an accident.' (p. 18, italics added). For other researchers, the solution has been to use different levels of culpability, creating different dependent variables. These authors, however, came up with the idea of weighting preventable and non-preventable accidents differently before adding them together. The formula used was

$$\frac{\text{Months worked}}{3 \times \text{no of preventable accidents} + 2 \times \text{no of non-preventable accidents}}$$

Thus, for reasons that were not stated, the authors regarded preventable collisions as 1.5 times more important than non-preventable ones. This would seem to indicate that what was being predicted was not really accidents, but how often and how severely drivers erred.

This kind of approach can also be found in the study by Kahneman, Ben-Ishai and Lotan (1973), where their dependent variable was a sum of the 'severity of error of the driver' in each crash. Unfortunately, this novel and potentially powerful development was not discussed or statistically investigated.

Similarly, a multiplicative approach based on responsibility was used by Corbett and Simon (1992) and Simon and Corbett (1996), where each collision was weighted by (self-reported) culpability on a five-point scale, divided by mileage and averaged over three years. No rationale was given for this method, apart from the statement 'In order to strengthen our criterion of accident rate ...' (Corbett and Simon, 1992, p. 61). However, it would seem to be very similar to the method of categorizing by culpability, as no-fault crashes were given a weight of zero.

Only one example of weighting traffic collisions for severity has been found;[11] Hartley and El Hassani (1994) included severity in their composite accident index. They also took recency of collision involvement into account, as did Schuster and Guilford (1962), and Schuster (1968). Note that weighing for severity is very different from weighing for recency; the first works under the assumption that worse drivers have worse accidents, the second tries to counter the diminishing validity for a given measure of behaviour for longer time periods.

One type of method, which could be called weighting, is to use the economical cost of collisions as criterion, a variable that would probably associate rather strongly with severity and number of injuries. This method is fairly unusual, but in Table 6.4 can be seen the secondary correlations for this variable, used by Carty, Stough and Gillespie (1998), comparing it to ten other outcome variables. It should be apparent that there is no pattern, and also that the variable that should be most similar, accident seriousness, instead yields a totally different pattern of effects, as the correlation between them and those of the costs is negative. In total, there is no useful information to be gleaned from this study, and whether the cost of collisions could be a useful individual differences outcome variable in traffic safety will have to be left unanswered.

---

11   For industrial accidents, Drake (1940) multiplicated number of accidents by their severity scores and divided it by length of service.

As a curiosity, a study by Ghosh and Tripathi (1965) on the accident proneness of factory workers can be described. Here, the ratio between the days absent due to accidents and the total number of days was used as criterion, meaning that both number and severity of mishaps were taken into account.

In conclusion, although a number of researchers have used various weighting methods for constructing driving safety criteria, none has discussed or even described their undertaking in any detail. If this type of method has any merit, it is still there to be discovered.

Mixed Criteria

> ... there is a temptation to devise programmes of research which study 'behaviour' (vehicle speeds, headways and so on) or 'driver errors' (however subjectively defined) because they are so much easier to 'observe' than accidents, without worrying too much about the relationship between accidents and observable behaviour. Maycock, 1985, p. 334.

As discussed previously, traffic safety can only be one thing; the absence of collisions. If there are no traffic accidents, then all other 'safety'-related variables and questions simply loose their importance. Therefore, the only possible measure of individual differences in traffic safety is collisions. However, many researchers are in the habit of using alternative criteria in their studies, something that will be considered in Chapter 7. Here, the in-between method of mixing collisions with other variables and using this amalgamation as an outcome variable will be discussed.

This subject is similar to the previous descriptions of criteria with weights applied, in terms of changes being made to the basic unit of accident as outcome. However, it is clearly a separate method that is worthy of a separate analysis, due to its logically very difficult nature. In principle, the term mixed criterion as used here is the method of taking number of collisions and adding (or subtracting, although this does not seem to be used) the number found on another variable (usually violations) for each individual, thereby creating a variable amalgamated from different sources. Many of the add-ons are what could be called alternatives to the accident as criterion. Others have no discernible (not to mention validated) connection with crashes. Most often, only one other variable is added, but there are a few examples of very complicated and thoroughly mixed criteria.

Accidents Plus Accidents

Starting gently, one possibility to create a criterion is to add together traffic accidents and other, non-traffic, mishaps. From an accident proneness perspective this would make sense, although the evidence for any generality of such a trait is sparse, and the correlations between different types of collisions usually are weak, as reviewed in Chapter 3.

Only two examples of a mixing of different accidents have been found. Junger, Terlouw and van der Heijden (1995) apparently asked respondents about any type of mishap, although the item formulation and the text of the paper were not in agreement about this. Similarly, Deffenbacher, Oetting, Lynch and Morris (1996), used an unspecified accident item that might have contained accidents from various avenues in life.

Turning thereafter to yet another mixed criterion, we finally find one that might have some sort of relevance and validity, although such an interpretation is guesswork, as no discussion of the method was found, as usual. Jelialian, Alday, Spirito, Rasile and Nobile (2000) used injuries acquired in motor vehicles as dependent variable, with the unusual definition of including these both when the person was driving and riding. From a driving perspective, this is of course very peculiar, but from a medical or wider social point of view this variable does have some merit. It is important to see whether people's injuries can be predicted. However, it could be claimed that a split into two variables could have increased the predictive power of the study, as well as the possibilities to interpret the results. Somewhat similarly, Beirness and Simpson (1988) formed two groups of subjects who had had an accident as vehicle passengers, either with a young or an older driver, and compared these with the traditional groups of drivers with and without collisions.

Accidents Plus Violations

The amalgamation of collisions and violations (traffic tickets) would seem to be the most common mixing method.[12] Given that the variance of violations will usually be much larger than that of crashes, it can be asked how useful this method is (no discussion of it has been found), as the amalgamated variable will mainly reflect the citation variable. This can be shown in a study by Cellar, Nelson and Yorke (2000), where the tickets/accident variable correlated .89 with tickets alone, while their effects correlated .94 (N=5). As a variation on the crashes plus violations formula, Wilson and Jonah (1988) added licence suspensions to these. A somewhat more advanced process was used by Donovan and Marlatt (1982); a

---

12   Used for example, by McGuire, 1956a; 1956b; Mozdzierz, Macchitelli, Planek and Lottman, 1975; Crancer and McMurray, 1968; Gutshall, Harper and Burke, 1968; Crancer and Quiring, 1969; Crancer and O'Neall, 1970; Fitten et al., 1995; Cellar, Nelson and Yorke, 2000; Richards, Deffenbacher and Rosén, 2002; Beck, Yan and Wang, 2007. Sometimes, the numbers are not mixed, but these variables used as group definers (for example, a group having both accidents and violations; see for example, Pelz and Krupat, 1974).

weighted risk index of collisions, injuries and violations. They gave no details[13] of the weighting process.

Accidents Plus Near Misses

The adding of (self-reported) collisions and near-accidents into one variable is also practiced by some researchers (for example, Gonzalez-Rothi, Foresman and Block, 1988; Engleman, Hirst and Douglas, 1997; Tirunahari, Zaidi, Sharma, Skurnick and Ashtyani, 2003; Pérez-Chada et al., 2005). Here, the example of Wu and Yan-Go (1996) demonstrates the outright flippancy shown by many researchers regarding their outcome variables. The amalgamation of two variables was only mentioned in the method section, where after the terms used were 'accident' and 'MVA' the latter spelled out as Motor Vehicle Accident, as if only an accident variable was used. In the end, it is therefore hard to know what these authors included in their criterion.

Yet another type of mixed criterion was used by Iversen and Rundmo (2002); crashes and near-accidents as driver or passenger. Adding collisions as passenger is mysterious, as no explanation of this was put forward, and no validation of this variable has been found in the literature. Furthermore, what use could predictive power versus this kind of incident have? Maybe there would be one person less injured in the collision, but this really needs to be tested before using this variable as a criterion. Again, it is hard to know what the authors were trying to do with their research, especially as they in their introduction pointed out; 'Mechanisms behind different types of accidents can differ, and these should be treated separately in analysis ...' (p. 1253). Apparently, they meant the predictor variables, as they did not extend their own advice to the outcome variable. A similar mixing was undertaken by these authors in their 2004 study; near accidents, collisions and accidents were someone was injured. At least, this time all these were for the respondent as driver. Instead, another oddity was introduced; the LISREL[14] model showed attitudes explaining risk behaviour, which explained the mixed criterion, which in its turn was used to explain its components (at least in terms of how the flowchart was drawn) of collisions, near accidents and injuries. Exactly how this should be interpreted was not described, but circularity would seem to be close. It could of course be a case of a blind acceptance of the statistical model outcomes,

13   An obscure reference was given for this method; Clay M. L. (1972). Which drunks should we dodge? In Selected Papers of the Twenty-Third Annual Meeting of the Alcohol and Drug Problems Association of North America. Washington D.C.: Alcohol and Drug Problems Association. This paper has not been possible to locate.

14   A statistical method that is fairly popular in the social sciences, mainly used to model data of a complexity that is way beyond any human understanding. The output is a series of weights for the associations of different sets of variables with each other. This is often shown as a flowchart with arrows between boxes.

regardless of whether the input variables were compatible with the assumptions of LISREL, or the output interpretable in any empirical (as opposed to mathematical) way, but, given the paucity of the text, this is impossible to know.

Accidents Plus Two

The next step on the mixed criterion ladder would seem to be collisions, violations and close calls. This recipe was used by Gilley et al. (1991). However, they also used accidents only as criterion in parallel. Unfortunately, the data reported were not suitable for a secondary correlation analysis, as several different scales were involved.

A mix with a very unusual ingredient was that of Marottoli, Cooney, Wagner, Doucette and Tinetti (1994), Marottoli and Richardson (1998) and Marottoli et al. (1998), where (self-reported) crashes and moving violations were added to being stopped by the police (apparently a milder form of violation variable). Even more unusual, however, was the inclusion in that paper of a short discussion of the choice of criterion; 'Because the current study was intended to develop an approach clinicians could use to raise their awareness of individuals who might be at increased risk for safety difficulties, the identification of even minor events was desired. Consequently ... we chose to define the outcome broadly ...' (p. 567). It is uncertain whether 'minor events' designated collisions or the actual events included in the criterion, and it is therefore not possible to really know how the authors were thinking about this. Either, they assumed that violations are a proxy for crashes, or they somehow thought that being stopped by the police and getting tickets were undesirable events in themselves, from a societal point of view. The use of the phrase 'safety difficulties' is similarly vague, and does not help in understanding the rationale of this study. It is important to note here that the peculiar criterion construction used by Marottoli et al. was not recognized as such by Anstey, Wood, Lord and Walker (2005), who in their review noted the very strong association in the 1998 study for one type of test with 'crash risk', despite that they had correctly described the mixed nature of the variable in a preceding table. In such a situation, the spreading of an urban myth is easily started, because few authors would seem to actually check the first-hand references, but instead rely on the restricted information given in second-hand sources, and sometimes forget the important information about the construction of the criterion that was actually there.

Yet another calculation method was invented by Gonzalez-Rothi, Foresman and Block (1988). Apparently, they mixed collisions with near misses and the discontinuation of driving (for fear of falling asleep). No interpretation of this combination is offered here.

Accidents Plus Tutti Frutti

One of the most profoundly mixed criteria found so far was used by Deffenbacher, Lynch, Filetti, Dahlen and Oetting (2003); moving violations, loss of concentration, minor loss of vehicular control, close calls, minor accidents and major accidents, all in one variable. No explanation was given to this cocktail of different ingredients, nor any validity coefficients. It can also be noted that in other studies, this research group has repeatedly stated that these indices do not form a reliable scale[15] (for example, Deffenbacher, Lynch, Oetting and Swaim, 2002; Deffenbacher, Richards, Filetti and Lynch, 2005). In the 2003 study, however, an alpha of 0.45 was reported, and the items amalgamated. The effects of this method were not discussed, but the different variables were also analysed separately.[16]

So far, the most advanced and cryptical amalgamational approach was used by Nagatsuka (1970) who computed criterion scores '... from the following formula: Coefficient of accident=Score of accident frequency+mean liability scores for accidents+score on skill and attitude in driving+ratio of risk in driving route+score for undesirable disease as drivers.' (p. 117). Although none of these five different concepts was explained in any way, a few interesting features can be discussed. In agreement with Moffie, Symmes and Milton (1952), Corbett and Simon (1992) and Simon and Corbett (1996), a sort of weighting of accident number by the degree of culpability of the driver seems to have been used, while the environmental risk of different routes was held constant, both of which, theoretically, are desirable features. However, how skill, attitude and 'undesirable diseases' could be brought together with the other scores is a mystery.

Grouping by Mixed Criterion

The construction of groups/samples defined by a mixed criterion can be seen as a methodology somewhat related to the previous ones described in this section, but with possibly different consequences. Here, a group is formed by taking, for example, drivers with one collision and many violations (for example, McGuire,

---

15   The reference to a scale would seem to indicate that these authors were thinking along psychometrical lines, believing that the responses on these items were similar to other scales, like for attitudes, in their properties. Although this is probably true in many ways (susceptibility to social desirability, for example), nevertheless this is not how accident item responses are usually seen. They are thought to correspond to actual, physical happenings in the outside world, not the private inside realm.

16   In general, this research group would seem to have a strong tendency to use un-validated, mixed outcomes variables, where the most blatant example is Deffenbacher, Oetting, Lynch and Morris (1996), where all criteria were apparently mixed scales containing items on actual behaviour and feelings. This included the extremely unspecified item 'Get into an accident'.

1956c; Haselkorn, Mueller and Rivara, 1998; Cox et al., 2003) and the characteristics of this sample compared with other groups' (for example, drivers with no crashes or violations). The result is that more restricted samples are used than if only collisions had been the criterion. Next, two basic possible consequences of such a method can be identified. The intention of this method would seem to be to form even more contrasted samples, under the assumption that the add-on variable adds a measure of safety that collisions cannot capture, which is one possibility. However, if the add-on does not carry any special information regarding safety, it will not add anything, but possibly introduce random error variance. Last, if violators have characteristics that are different from those of accident-prone drivers, then a systematic bias will result. As usual, this method would seem to never have been discussed by those who use it (for example, Harano and Peck, 1972).

Concluding Remarks

The basic problem of all the mixed criterion methods is that, for reasons unknown, the authors seem to believe that collisions and other variables are equivalent, or at least strongly correlated. That this is most often not the case will be shown in Chapter 7. The consequence of this mixing of lemons and oranges (perhaps not very different, but definitely not the same) is that it is hard to exactly pinpoint where the significant effect resides, although it can be suspected that it is usually not with the collision component.

Various Odd Methods

Differences Over Time

> Moreover, the potential risk for unsafe motor vehicle operation may be underestimated by the reliance on accident rates alone. Gilley et al., 1991, p. 945.

The possibly most intriguing way of calculating a criterion in traffic accident research seems to have been invented by Bingham and Shope (2005). At least, no one else seems to have used it, and no reference or explanation was included. First, all crashes for a variable time period were used, although the variation was not large between subjects, as they were all in their twenties. Second, this period was divided into one before and one after their twentieth birthday, and the dichotomised values of these two compared for each subject. In this way, four groups could be constructed; 'Stable Low', 'Stable High', 'Increasing' and

'Decreasing'. Apparently, for reasons untold,[17] the authors thought that the change in numbers between periods was significant, that is that some kind of trend was detected in this way.

Although fascinating in its very different approach, the change-over-time principle would seem to run into several problems, logical and statistical. For example, the uncommonness of traffic collisions would make this kind of categorization very unreliable, and it is uncertain if the 'Increasing' group is in for some kind of infinite increase. Finally, it can be observed that the significant differences found may simply have been due to different means in the groups, as this was apparently not tested.

Different Data Sources

In a study by Eadington and Frier (1989), on diabetic drivers, the self-reported accident rate of these patients was compared with recorded rate for the general population. As self-reports usually yield a higher count than records, for the same drivers, this method cannot be said to have any validity. Fortunately, these results probably erred on the safe side, yielding no difference. In reality, the diabetic drivers probably had a lower collision rate than the general population, unless their mileage estimates were biased towards over-reporting. In the end, no conclusions at all can be drawn about diabetic drivers from this study, because there were too many unknown biasing factors whose effects cannot be estimated. The only conclusion is that if unreliable sources of data are to be used in comparisons between groups, these should be used for all of them.

Discussion

> ... it is worth bearing in mind that accident frequencies ... are the measure of success or failure of any road safety measure. Maycock, 1985, p. 334.

In the review of the published papers for this chapter, a bewildering array of safety criteria has been found. No consensus and hardly any discussion are to be found in the literature. Actually, what few sentences most often describe the outcome variables are in many cases almost bereft of any useful information about how the data was gathered. In Hartman and Rawson (1992), for example, not a single word was said about how the collision variable was constructed. In fact, it was only stated in the Discussion that these were 'automobile accidents'.

In general, this problem can be described as due to the varying interest of researchers, where often meticulous care has gone into the measurement and discussion of the predictors and very little has been left for the predicted variable,

---

17   It is possible that the reason lies with Problem Behavior Theory, which was given as a general framework for the study.

which is often only mentioned in a few sentences (for example, Iversen and Rundmo, 2004). The total lack of interest in the collision variable is astonishing, especially given its importance for improving traffic safety.

It can therefore be said that the reasons for the various criterion models reported on here have never been stated with sufficient clarity for a full understanding of these undertakings. Neither has any research been undertaken with the aim of clarifying what possible differences can result when different criteria are used. Most researchers have tested different predictors against the same criterion to see which one is the most powerful, instead of trying different criteria against the same predictor. This means two things:

1. We do not know which way of constructing an accident criterion is superior.
2. We do not know whether any single result in collision prediction studies is due to the power of the independent or the dependent variable.

Furthermore, as some researchers have used mixtures of different variables when constructing their safety criterion, we sometimes do not even know what has been predicted. The information gleaned from such composites would therefore seem to be very limited. No argument in favour of this method has been found.

There are also instances of methods that do not even allow categorization under any of the many section headings used here. King and Clark (1962), for example, took the highest number of crashes for each driver that could be found in state records and by self-report. How is such a mix of sources to be interpreted?

Finally, it can be noted that the other side of the criterion coin would seem to be what statistical method is used. Here, some kind of comparative research would seem to be needed, especially for the odds ratio method, which is growing more and more popular. It would be interesting to know how an odds ratio translates into a correlation size. From what little evidence there is, it would seem like the odds ratio is a researcher's trick to make an effect appear more important than it is, as the odds ratio is a relative measure of risk, not an absolute, like correlations and regressions. If the initial risk is very small, the resulting odds may be very high, despite the actual association being too weak to be of any practical importance. For the same reason, the odds ratio probably often yields different results to other measures, in terms of relative importance of variables.

The lack of comparative studies and theoretical statements on the topic of safety criteria make it hard to draw any more conclusions than what has been stated here, or even discuss the subject. More research is definitely needed. However, the most important thing needed is that researchers stop using mixed and weighted criteria until we know more about what this actually leads to in terms of effects on results and conclusions.

Given the many peculiar methods described in this chapter, it can, for reasons of probability only, be strongly suspected that a large share of studies on individual differences in accident liability are worthless, because they have predicted variables that have little to do with safety. All these different versions cannot be correct.

Chapter 7
Alternatives to Accidents as Dependent Variable

To be successful in avoiding accidents would seem, theoretically speaking, the ideal criterion for successful driving performance. However, accidents are rare events, and studies of traffic accidents are therefore, from a statistical point of view, a weak instrument for acquiring evidence of driver behaviour. Hence, surrogate measures for 'success as driver' must be used. Rimmö and Hakamies-Blomqvist, 2002, p. 48.

Introduction

... a theory that can predict a person's proneness to engage in risky driving behavior but cannot predict the ultimate criterion of traffic accident casualties would not be very useful. Wilson and Jonah, 1988, p. 176.

One conclusion that can be drawn from the previous chapters is that traffic accidents are extremely hard to use as a dependent variable, due to the many methodological and statistical problems associated with them. We do not even really know how to construct the accident variable in an optimal way when using it as a criterion. To summarize, characteristics of collisions include:

1. Being uncommon, which means that statistical power is low.
2. Hard to define, in the sense that what should be actually considered an incident eligible for inclusion as criterion differs with the goal of the study, resulting in extra error variance.
3. Quality of data often being very poor, as no really reliable sources exist, possibly apart from some company records.
4. Reliability is low for those populations and time frames that are commonly used in traffic research.

Some researchers have therefore concluded that it is better to use other driving-associated variables as proxies for accidents, as it is then possible to gather data that do not share the shortcomings listed above. Actually, using something else than accidents as the criterion of safety has become rather common.[1] The basic

---

1   For example, Adams-Guppy and Guppy, 1995; Wu and Yan-Go, 1996; Heikkilä, Turkka, Korpelainen, Kallanranta and Summala, 1998; Beck, Shattuck and Raleigh, 2001; Machin and De Souza, 2004; Risser, Chaloupka, Grundler, Sommer, Häusler and Kaufmann, 2008; see also all tables in this chapter.

assumption[2] of proxy variables (in individual level prediction studies) would seem to be that some kind of incident or behaviour is strongly associated with accidents and can therefore be used as a dependent variable instead of crashes, with the advantages of being more common, and that the quality of the data is restricted mainly by the quality of the data gathering undertaken. It is also assumed that proxy variables will yield the same relative result as an accident variable with very good validity. If those assumptions were actually true, the only advantage of accidents as a dependent variable would seem to be that large databases are already in existence and can be used with very little extra work.

However, as will be shown, the use of proxy variables share the same type of problem as many of the other traffic variables discussed throughout this book, that is its validity is assumed but not tested. Instead, proxy variables are most often used in traffic safety studies without this being explicitly acknowledged or even discussed by the authors.[3] When discussed, the validity of the proxy variable used is often poorly referenced (for example, Consiglio, Driscoll, Witte and Berg, 2003), and/or based upon erroneous logic, as will be discussed in the next section. Instead, we are expected to believe and accept at face validity that certain behaviours are indeed dangerous, to such a degree as to differentiate between drivers in terms of safety, without this having been proven. Often, this acceptance comes easy, as most of us seem to have an innate conception of dangerousness, which makes us think that certain acts are not safe. However, in science, this is not a good criterion, and we should instead, as usual, look for the actual evidence.

Another facet of the proxy problem is that some authors do not seem to make a real distinction between the actual safety variable of accidents and various alternatives when discussing other research. Often, they are seen as the same thing, as exemplified in Chapter 6 regarding mixed criteria. Actually, some researchers even seem to think that accidents are somehow inferior as a criterion variable (for example, Rimmö and Hakamies-Blomqvist, 2002).

The present chapter takes a look at some of the many proxy variables used by various researchers, discusses the logic behind them (if any can be discerned), and relate what evidence there is for their validity. When analysing the reported evidence, there are two main ways of doing this. Rather self-evidently, the within-individuals association between a proxy and accidents is of prime importance. However, given the difficulty of finding strong associations with accidents in the first place, it can be expected that these too will be weak, and it can always be argued that this is due to the normal problems of using the accident variable (summarized above), which leads to the use of proxies in the first place. Therefore,

---

2   Actually, these assumptions are hardly ever discussed, but they are necessary for the use of proxy safety variables. If they are not made, it is directly admitted that something else than safety is predicted.

3   For example, Heath, 1959; Donovan, Queisser, Umlauf and Salzberg, 1986; Harré, Brandt and Dawe, 2000; Hartos, Eitel and Simons-Morton, 2002; Beck, Wang and Mitchell, 2006; Fernandes, Job and Hatfield, 2007.

the secondary correlation method described in the introduction will be used whenever possible. The logic behind this method in the present context is that if a proxy is a valid replacement for accidents as dependent measure, it should yield a similar pattern of associations with other variables, otherwise it will lead to erroneous conclusions. Also, it is expected that the associations with predictors for proxies will generally be stronger, due to higher variance. If not, there seems to be precious little meaning in using them.

Two basic types of proxy variables can be identified; incident and behaviour parameters. The first consists of happenings that are experienced by drivers as pre-stages to accidents that are stopped in time, or incidents where road users have come extremely close to each other in time and space. The second type is those behaviours that are considered dangerous, that is might lead to accidents although they usually do not. Finally, a few proxy variables that are a bit hard to define will be discussed. First, however, an analysis of the most common arguments for proxy use will be undertaken.

Alternative Logic

> Several authors ... outlined that accidents can be partially attributed to factors unrelated to unsafe driving or other individual factors. Sommer, Herle, Häusler, Risser, Schützhofer and Chaloupka, 2008, p. 363.

The use of alternative outcome variables in research that is ostensibly about (individual) traffic safety is so widespread and accepted that a challenge to this habit apparently first must make visible the various assumptions made about these parameters. Here, two basic modes of argument can be discerned; negative (regarding accidents as outcome) and positive (regarding the qualities of the replacement variable). The latter contains a subgroup, where it is argued that the variable used is important for accident causation, and therefore a good proxy. This argument will be studied first, as it is logically different from the other positive arguments, although this does not seem to be realized by the researchers using it.

Intra Versus Inter-Individual Effects

Why have safety proxy variables, like speed, come to be accepted as valid replacements for accidents in individual differences in safety research? Here, it will be argued that one reason is that evidence from within-individuals studies have been applied to between-individuals (individual differences) research.

For example, it is well known that the number and severity of accidents on a road section or other geographical area are influenced by average speed and speed limits (Farmer, Retting and Lund, 1999; for a review, see Aarts and van Schagen, 2006). This is a within individuals effect, that is what is (mainly) measured is a difference in speed for driver A, on different occasions. What can be claimed from such data is that if driver A reduces his speed, he will with great certainty be a safer driver (unless the pure physical differences of less energy etc is offset by a change in vigilance), because all other things are equal for this driver. Such a within-individuals difference is not contested here, especially not at an aggregated level. When speed is measured for different drivers, however, the situation is very different. Even if driver A is driving much faster than B, this is not conclusive evidence that he is the more dangerous driver, because driver B might be causing his crashes by other unsafe behaviours.

The problem is that the proxy variable speed as used by many researchers as an outcome parameter is applied on between individual effects, and at this level, there is not much overlap with accidents. Judging from the studies in Table 7.5, where such data is summarized, even a very favourable interpretation would indicate that less than a quarter of the variance is shared. Therefore, studies using speed as a proxy could in principle be three-quarters wrong, or more.

That many researchers are confusing within- and between-individual effects can be seen by how they use references for both these when arguing for their proxy use (for example, Brown and Cotton, 2003). Sometimes, no references at all are given, and no difference apparently made between these two very different effects (for example, Åberg and Wallén Warner, 2008).

Base Rate Neglect[4]

Another type of logical confusion in the argumentation for proxy use is when accident analysis results are applied at the individual level. For example, it has been said that 'An important consideration in the selection of laboratory tests is the observation that errors of attention are the most frequently cited errors leading to actual traffic accidents involving ethanol.' (Gengo and Manning, 1990, p. 1036). The logic behind such a statement would seem to be that if a certain variable is often found in crash investigations, it should be a good predictor of (individual differences in) accident record, and/or a good replacement for it as criterion. However, a certain occurrence being present in the unfolding of an accident does not necessarily make it a good predictor, because that does not take into account how common the phenomenon is. This could be illustrated by the case of breathing; it can be stated with great certainty that almost all drivers involved in accidents were breathing at the time of the crash (it could even be claimed that this was a part of the cause of the crash). However, as we are all breathing all the time, the predictive power of this variable is therefore zero.

This kind of proxy thinking has probably been imported from safety intervention, where prediction, in terms of individual differences, is not really an

---

4   Base rate neglect is a cognitive phenomenon where people ignore how common a thing is when estimating the probability of occurrence (Kahneman and Tversky, 1972).

issue, but overall methods are used. Thus, all drivers are forbidden to drive under the influence, which is probably a good intervention, despite that it is not a very good predictor for individual differences in accident record (meaning that most drunk driving does not result in accidents). This is a case where raw power of influence is used instead of predictive power. Speed is also such a case, where the safety effect is reached by restricting everyone's speed, not just those who cannot handle speed in a safe way.

Arguments for Proxy Use

> Moreover, accidents are not useful at all for behavioral analysis. One cannot expect to witness an accident during a short driving test, much less to be able to describe the event and to investigate the causes of it. Besides, if accidents were to be expected during a driving test, it would certainly be difficult to find observers! Risser, 1985, p. 180.

Why does the traffic research community accept the use of un-validated proxy variables? As stated before, in most instances of papers that use such dependent measures, the choice of parameters is not discussed at all, but simply given (for example, Alm and Nilsson, 1994). This behaviour by authors can be interpreted in two ways; either they have such confidence in their methodology (and think others have too) that they think it is superfluous to discuss it, or they are consciously glossing over a part of their research that they know is questionable.

In the present section, what few arguments regarding proxy use that have been found will be discussed. These have some similarities to the techniques discussed regarding self-reports in Chapter 2. However, there are also some interesting differences, mainly regarding how the authors try to argue for a concept of safety that does not include accidents. The arguments here come in two forms; the negative, where it is claimed that the unreliability of accident record, and other methodological difficulties, make it unsuitable as dependent variable, and the positive, where the argument is that the proxy is important for safety.

The negative argument is basically the same as the one used regarding self-reports versus records, and suffers from the same logical error; showing that one variable is a bad criterion does not automatically make another one better. The basic negative argument can therefore be dismissed as fairly primitive. However, there is also a claim that is sometimes seen, which seems to be a further development of the negative argument; the proposal that accidents are not a good measure of risky driving (for example, Ulleberg and Rundmo, 2002), an idea that seems to go beyond the statistical and record-keeping properties of this variable. As stated by Gilley et al. (1991); 'Moreover, the potential risk for unsafe motor vehicle operation may be underestimated by the reliance on accident rates alone.' and '... the tacit equation of accident rates and other forms of unsafe driving with deficient "competence" or fitness to drive has face validity but oversimplifies the situation.' (p. 945). How 'unsafe' could be defined or measured without referring to accidents was not explained. Anyway, this claim assumes that 'risky driving' exists as something that is independent of and independently important from traffic accidents. These authors therefore make the mistake of believing that we can in some way ascertain what is 'risky' without referring to collisions, but they do so without saying this explicitly, or giving any evidence. The negative argument is therefore rather hollow, and easy to dismiss, when it has been analysed.

The positive argument is trickier to handle, mainly because it is usually even more vaguely stated than the negative. The illogical type uses the irrelevant results discussed above, discussing how many accidents are caused by a certain type of behaviour (for example, speed), and thereafter move on to study what influences this variable, returning to safety in the discussion section. Here, it can also be noted that the argument in favour of a certain variable as an important cause of accidents (and therefore a good proxy) is thereafter often turned on its head, when it is claimed (often as an explanation of the weak results) that 'accidents have many causes', and that the association therefore cannot be very strong (for example, Ulleberg and Rundmo, 2002).

Actually, by observation, it could be claimed that all of the 'risky' behaviours that are commonly discussed within traffic safety are instead very safe, because we would hardly ever see a collision while observing. In the end, all we have as evidence concerning the risk of various behaviours, if we do not want to associate them with crashes, are our gut feelings, and these are certainly not a very good guide towards anything even resembling scientific knowledge. Sometimes it does not even lead us to a good lunch.

The other common form of the positive argument starts out by pretending that the problem studied, for example, the 'attitude-driving behaviour relationship', is something worthy of interest in itself, which makes the choice of dependent variable natural. However, in the end the issue of safety crops up anyway, and it is acknowledged that the study was about (individual) traffic risk. By the time this assumption is inserted, the reader will probably have forgotten that safety was not given as a reason to begin with, and that no evidence was given regarding validity. This method for avoiding unwanted results has, for example, been practised by DePasquale, Geller, Clarke and Littleton (2001), who developed a scale for Propensity for Angry Driving (PADS), where driving scenarios are described ('someone makes an obscene gesture at you') and various possible reactions given as alternative reactions ('a) Ignore the driver'). Most interestingly, the authors of this study said that 'The PADS was constructed to assess an individual's propensity to become angry with other drivers. Therefore, the PADS score should predict the occurrence of angry or hostile behaviors performed while driving.' (p. 7). In line with this reasoning, they used self-reported confrontations with other drivers as criterion for the PADS.

Now, one can ponder what the logic behind this method was. For some reason, the authors wanted to predict the propensity to become angry with other drivers. But if the number of self-reported such occurrences was seen as an acceptable criterion, why then construct the PADS in the first place? Why not just ask the drivers about their confrontations with other drivers, if that was all the scale was about? If it was actually about driving safety, no evidence was presented for it. As with so many other studies, face validity was all that was offered as evidence.[5]

There is also a similar but somewhat more advanced variation of the positive argument. Some papers present a fairly logical reason as to why a certain parameter is used, for example, cognitive interference is expected in a certain situation, and this is measured as differences in control of the vehicle (Matthews, Dorn, Hoyes, Davies, Glendon and Taylor, 1998). This set-up looks deceptively straightforward and valid; the hypothesis points directly to what variables should be used.

However, the question that should be asked is; why is the study undertaken in the first place? All the research referenced in this chapter had, to some degree at least, safety as one of their goals or reasons, although this was far from always explicitly stated. But what is safety? Is it low speed, or little lateral speed change, or hazard detection, or what? Actually, safety is a lack of accidents, and all the variables discussed in this chapter are only behaviours that we believe, sometimes rightly, to be related to safety/accidents.

So, a disturbing amount of traffic 'safety' researchers use our implicit understanding of what safety is, and is not, to publish studies that contain highly dubious outcome variables instead of the real thing. In the words of Goldenbeld, Twisk and de Craen (2004): 'Given the great difficulties in establishing the effects of practical training on accident rates ... we chose to assess the effects of training on riding competence as the next best feasible option.' (p. 2). In other words; if you cannot show an effect on safety, show an effect on something else, and then claim it is 'safety-related'. Having analysed what proxy researchers say about their variables, we now turn to a part where they are even less talkative; the evidence.

Incident Proxy Variables

> If these ... are not correlated with the ultimate criterion of accidents ... it is said
> to lack validity. McGuire and Kersh, 1969, p. 19.

There are several differently named traffic incident variables which have been used as replacements for crashes, but they all seem to be rather similar; 'traffic conflicts' (for example, Campbell and King, 1970; Chin and Quek, 1997), 'unsafe driving incidents' (McKnight and McKnight, 1999), 'close calls' (Deffenbacher, Filetti, Richards, Lynch and Oetting, 2003), and 'near accidents' (Underwood, Chapman, Wright and Crundall, 1999; Iversen and Rundmo, 2002), and if any distinction is to be made, it should be between the first and all the other.

---

5   However, other authors have included this scale in their studies and tested it against (self-reported) accidents; Dula and Ballard, 2003; Dahlen and Ragan, 2004; Dahlen, Martin, Ragan and Kuhlman, 2005.

Traffic conflicts is, in comparison to the other incident variables, a fairly well discussed and somewhat formalized variable (for example, Hauer, 1982), mainly used by traffic technicians for safety evaluations of places, normally junctions. Observers gather data. Close calls and similar concepts, on the other hand, do not seem to have any explicit definition, and rely on the individual driver as reporter of data (and also as interpreter of the concept, as a definition is usually not included in the questionnaire). The basic idea is probably the same as in traffic conflicts; the incident is an unforeseen traffic event where some party needs to take evasive action to avoid a collision, and succeeds in this.

Close Calls/Near Misses/Near Accidents

> Frequent spontaneous comments were given ... from patients with sleep spells, describing how they had experienced several near misses, which further indicates the hazard in traffic. Haraldsson, Carenfelt, Diderichsen, Nygren and Tingvall, 1990, p. 61.

There seems to be widespread belief within the research community in the validity of close calls as a proxy variable for accidents.[6] For example, Barger et al. (2005) said that the effects of their predictors on health and safety '... has not been evaluated with the use of validated measures.' (p. 125). As this was written in the introduction, it can be presumed that the authors thought they were doing this themselves. One of the variables used was near-miss incidents, for which no reference was given, and the subject of its validity not even discussed. Given that this is how incident proxy variables are usually used (for example, Deffenbacher, Lynch, Filetti, Dahlen and Oetting, 2003), one can ponder what the validity of near misses as a road traffic safety measure actually is.

First, it can be noted that the researchers using near misses are remarkably silent about its validity as a safety proxy variable. The Deffenbacher group, for example, never seems to have discussed the problem or given any references (see references in Tables 7.1–7.2), despite their prodigious use of it.

A few researchers have mentioned the validity problem of near misses, however. For example, Smith and Heckert (1998), when using personality measures to predict 'traffic accidents' (actually near misses) cited Edwards and Hahn (1980) for claims of the validity of near accidents. However, at least two circumstances of the latter study make its relevance for the Smith and Heckert piece doubtful. First, it was an industrial accidents study, and second, it was some sort of group level calculation. No association was reported for individual level data.

---

6   For example, Forbes, 1957; McFarland and Moore, 1957; Smith and Heckert, 1998; Häkkänen and Summala, 2000; Levelt and Rappange, 2000; Iversen and Rundmo, 2002; 2004; Harrison, 2004; Philip, 2005; Tzamalouka, Papadakaki and Chliaoutakis, 2005; McEvoy, Stevenson and Woodward, 2006.

As described in the introduction, there are two main ways of validating near misses as a proxy variable for individual accident record; by correlating with actual accidents at the individual level, and by using the secondary correlation method, that is checking whether the pattern of effects are similar for the two variables when they are used in parallel. Unfortunately, the question of near misses as a replacement variable has not received any special attention by researchers, and the data gathered and analysed here are from calculations undertaken without any explicit agenda for the analysis. Instead, it seems like many researchers have simply been conscientious in their reporting, including some results that were not thought to be important for their actual research questions. The only example found where the explicit aim of a study has been to validate close calls against collisions is Roberts, Chapman and Underwood (2004).

In Table 7.1 correlations between near misses and accidents reported in some studies can be seen. It would seem to be obvious that the association is very weak, despite the use of self-reported data, which probably inflates the association.

Apart from the associations in Table 7.1, Dahlen, Martin, Ragan and Kuhlman (2005) reported that 'None of the variables included in the present study predicted minor or major accidents.' (p. 345), which presumably included close calls, as those were among the variables used. Given the sample size and an assumed $p<.05$, this would mean that the correlation was less than .11. Why the results for accidents were omitted from the table over zero-order intercorrelations of study variables was not explained.

Turning to secondary correlations, a few studies have been found where this method could be utilized. Most of these used a division of accidents into major/minor, and are summarized in Table 7.2. The results ranged from the fairly strongly positive to almost as strongly negative, despite the use of self-reported data in almost all studies. It can therefore be concluded that close calls seem to be predicted by other variables than accidents.

However, one positive result has also been found. Aldrich (1989) used (self-reported) accidents overall, accidents due to sleepiness (the subject of the study) and near misses due to sleepiness as parallel outcome variables. The mean number of such incidents in five different groups (one such grouping for men and women, respectively, that is $N=10$) were (secondary) correlated. The sleep-related parameters correlated .90, but both were very weakly associated with overall accidents. Actually, for the female group, the latter associations were negative, and although sleep-related accidents should be the best criterion for a sleepiness study, this is a bit intriguing. However, it should be pointed out here that the similarity of means between groups is not the same thing as a similarity in effects, or a strong association at the individual level.

In similar areas of research, similar results can be reported; Marsch and Kendrick (2000) found that near misses did not predict future minor injuries of children. Likewise, Smith, Silverman, Heckert, Brodke, Hayes and Mattimore (2001) reported correlations of .06 and .14 between near injury events and recorded

Table 7.1   The association between close calls and accidents in various studies

Study	Accident variable	N	Statistic	Effect	Accident data source	Time period accidents	Time period close calls	Comments
Adams-Guppy and Guppy, 2003	All accidents	640	r	.237	Self-report	Variable	3 months	Very diverse sample. Accident variable dichotomized
Corbett and Simon, 1992	All accidents	471?	Not reported	ns	Self-report	Three years	Variable/ maximum 3 years	Reporting unclear. Near misses only when committing traffic offences
Dahlen and Ragan, 2004	Minor accidents Major accidents	232	r	.15 -.01	Self-report	Variable	3 months	Minor and major correlated .47. Close calls and accidents reported over different time frames
Dahlen and White, 2006	Minor accidents Major accidents	312	r	.20 .10	Self-report	Variable	3 months	Minor and major correlated .42
Deffenbacher, Lynch, Oetting and Swaim, 2002	Minor accidents Major accidents	290	r	.19 .03	Self-report	3 months	3 months	–
Morrow and Crum, 2004	All reportable and chargeable	116	r	.10	Self-report	2 years	2 years	Close calls used a ranked scale, and were only those due to fatigue
Roberts, Chapman and Underwood, 2004	All accidents	98	r	.273	(Self-report)	3 years	2 weeks	Driving diaries for both variables
Smith and Heckert, 1998	All accidents	76	r	.07	Self-report	Variable	Variable	Participants were students
af Wåhlberg, Dorn and Kline, forthcoming	All accidents	115	r	.00 -.06	Self-report Company records	Variable 2 years	1 month	Pick-up truck drivers

Note: Shown are the accident categories used, number of subjects in the study, type of statistic reported, the effect, source of accident data, and the time periods used for gathering the data.

Table 7.2   Correlations between effect sizes for close calls and accident variables (minor and major crashes)

Study	Close call/ minor crashes	Close call/ major crashes	Minor/major crashes	N	Original N	Mean effect minor/major accidents	Mean effect close calls	Comments
Dahlen and Ragan, 2004	.43	.11	.90	12	242	.209 (minor) .149 (major)	.210	Values from table 2
Dahlen and White, 2006	.69	.30	.78	11	312	.092 (minor) .075 (major)	.209	Values from table 2
Deffenbacher, Lynch, Oetting and Swaim, 2002	-.65	-.55	.05	7	290	.023 (minor) .101 (major)	.056	Values from table 4
Deffenbacher, White and Lynch, 2004	.77 (men) .62 (women)	.27 (men) .36 (women)	.19 (men) -.05 (women)	14	218 218	.081 (minor, men) .068 (minor, women) .102 (major, men) .073 (major, women)	.146 (men) .183 (women)	Values from table VI. Lifetime reporting?

Note: All variables were self-reported. All original statistics probably Pearson correlations. Shown are the associations between the original correlations for each variable with various predictors, the number of such predictors, original number of subjects, and the means of the original effects for the accident and close calls variables.

ones, and similar values for unreported injury events and recorded. However, they blamed this result on the 'widespread inaccuracy' of the records.

In general, it can therefore be said that accidents and close calls seem to be very different variables. If this result is some sort of artifact due to methodological shortcomings cannot be ascertained today, as few studies using objective data have been found. As with so many other aspects of driver behaviour, the technological advances in vehicle measurement will probably change this state of affairs, as it will be possible to actually measure some types of near misses.

Behaviour Proxy Variables

> The behavior cited in violation statistics is, for the most part, under the volition of a driver despite the probabilistic nature of detection. In comparison, crash involvement is less dependent on the behavior of a particular driver and more dependent on the environmental circumstances and the behavior of other drivers. Struckman-Johnson, Lund, Williams and Osborne, 1989, p. 204.

Behaviour that is not specifically tied to incidents is often used as a proxy for crashes. The difference as compared to incident proxies is mainly the level of danger; although a certain behaviour may in general be perceived as dangerous, it does not carry with it the same feeling of imminent risk as related to a specific object (usually another vehicle) and/or evasive action.

The behaviour proxy safety variables are actually often used as predictors of accidents, sometimes tested at the individual level of analysis, sometimes not. Speed is probably the most common example (sometimes conceptualized as 'speeding', that is breaking the law) of an easily measured variable that is accepted as an accident proxy.

However, there are also a number of other proxy variables used in various studies, which may seem to have the desirable properties of risk. But do they? Here, a number of examples will be scrutinized regarding the evidence for the variables' validity as replacements for crash variables. As with the other proxies, the main tool is association with accidents and secondary correlations between effects. It will also be asked whether the authors using the variable give any reference or evidence concerning its legitimate use as a proxy for accidents. In general, it is suspected that many proxy variables only have face validity, and that the authors have no real reason for using them, apart from a wish to have easily gathered data and significant results.

Citations/Violations[7]

> Violations, however, are correlated with accidents only to the extent of between
> .10 and .20 – a figure that may be statistically significant but in this case lacking
> much practical significance. McGuire and Kersh, 1969, p. 25.

In many countries, traffic tickets are used as a proxy for accidents by the state, because they are given for law breaking behaviour that is thought to be dangerous. This thinking has also been accepted by researchers in individual differences, who often use number of violations on record as a dependent variable (for example, Williams, Henderson and Mills, 1974; Brown, 1976; Hansotia and Broste, 1991; Burns and Wilde, 1995), or as a grouping variable (for example, Beamish and Malfetti, 1962; Quenault, 1967; 1968; Jamison and McGlothlin, 1973; Haselkorn, Mueller and Rivara, 1998), sorting people into samples by their citations.

It can be pointed out that violations are actually twice removed from accidents, because the tickets (or points) are given for behaviour that is considered dangerous, and as state recorded violations are probably a poor estimate of the actual number of these behaviours, which are in themselves probably poorly related to safety on the individual level (see the next section regarding speed), the end result must by necessity be very weakly associated with crashes at the individual level, although they may have an effect as a deterrent. As with so many other variables, having an effect at the group level and for individuals is not the same thing.

However, there is one methodological problem that may cause very strong correlations between traffic tickets and accidents; the former can in some countries be issued for the latter, thus causing an inflated association that is not really relevant, because there is total information overlap between sources. Unfortunately, far from all researchers have stated whether they have included fines issued for crashes in their violation variable (for example, Lourens, Vissers and Jessurun, 1999), and the evidence is therefore somewhat difficult to evaluate. That a sizeable problem could exist is indicated by the correlation of .46 between recorded accidents and 'colliding' violations reported by Edwards, Hahn and Fleishman (1977). As can be seen in Table 7.3, this correlation is one of the highest ever found.

Anyway, the evidence regarding the association between violations and accidents is rather plentiful, and especially the very large studies carried out in California, mainly by Gebers and Peck, deserve credit. As can be seen in Table 7.3, the correlations found have usually been very small. To these results can be added those of Cooper (1997), one of the few to state that tickets issued for crashes were

---

7   The use of the word 'violation' in the current situation should not be confused with its other uses, for example, within the Driver Behaviour Questionnaire, where it denotes various bad driving behaviours that are not necessarily illegal, or detected by the police. This double use of the term has apparently caused some confusion, and some researchers have equated the terms (for example, Lourens, Vissers and Jessurun, 1999). In the present chapter, 'violation' means having been legally punished for an infringement of traffic law.

Table 7.3   The association between traffic accident record and traffic violation record in various studies

Study	Statistic	Effect	N	Accident source	Predictors	Violation source	Comments
Arthur and Doverspike, 2001	Pearson?	.48	48	Self-report	Moving violations	Self-report	–
Burg, 1967	Pearson	.292	9,561	State record	Convictions	State record	Data from table 11
Burns and Wilde, 1995	Pearson?	-.05 .29 .12	51	State record	Speeding violations All violations except speed Demerit points	State record State record State record	–
Cellar, Nelson and Yorke, 2000	Pearson?	.30 .19	202	Self-report	Driving tickets	Self-report	The time period was given as 10 years, but as the sample had a mean age of 20.9, this time period cannot have been used for all
Dahlen and Ragan, 2004	Pearson?	.35 .35	232	Self-report	Moving violations	Self-report	Lifetime reporting
Daigneault, Joly and Frigon, 2002	r	.10252 .09715 .09720 .08129	187,620 131,334 71,637 35,818	State record	Convictions	State record	Non-concurrent prediction
Deffenbacher, Lynch, Oetting and Swaim, 2002	Correlation	.05 -.03	290	Self-report	Moving violations	Self-report	–
Edwards, Hahn and Fleishman, 1977	Correlation	.28	152	State record	All violations	State record	'Colliding' violations included. Violations probably for a variable time period
Ferdun, Peck and Coppin, 1967	Pearson?	.146 (males) .155 (females)	3,385 2,255	State record	Violations	State record	Drivers aged 16–19 1/2
Fergenson, 1971	phi	.196	15	Self-report	All	Self-report	High violation/low accident group had highest information processing scores

Table 7.3   *Continued*

Study	Statistic	Effect	N	Accident source	Predictors	Violation source	Comments
Furnham and Saipe, 1993	Correlation	.20	73	Self-report	Traffic convictions apart from alcohol related ones	Self-report	—
Galovski and Blanchard, 2002	Pearson	.036 (raw) .42 (years of driving constant)	30	Self-report	Moving violations	Self-report	Effect calculated from raw data in table 2. Removing outliers increased the correlation
Gebers and Peck, 1994	Pearson	.098 .080 .095 .096 .105	160,525	State record	Total citations Total citations excluding TVS Total countable citations Moving citations Negligent-Operator points	State record	Prediction of subsequent accidents from three years of violations
Gebers and Peck, 1994	Pearson	.120 .101 .115 .117 .127	152,931	State record	Total citations Total citations excluding TVS Total countable citations Moving citations Negligent-Operator points	State record	Prediction of subsequent accidents from three years of violations
Gebers and Peck, 1994	Pearson	.143 .131 .140 .140 .150	139,485	State record	Total citations Total citations excluding TVS Total countable citations Moving citations Negligent-Operator points	State record	Prediction of subsequent accidents from five years of violations
Gebers and Peck, 1994	Pearson	.162 .157 .160 .161 .173	114,618	State record	Total citations Total citations excluding TVS Total countable citations Moving citations Negligent-Operator points	State record	—

Table 7.3 *Continued*

Study	Statistic	Effect	N	Accident source	Predictors	Violation source	Comments
Goldstein and Mosel, 1958	Pearson?	.49 .68 .47 .55	323	Self-report	Moving violations for which fined?	Self-report	Lifetime reporting. Possible shared variance due to the fines being issued for accidents
Gresset and Meyer, 1994a	OR	2.41	4,036	State record	Demerit points	State record	Case-control
Knee, Neighbors and Vietor, 2001	Pearson?	.45	>99	Self-report	Traffic citations excluding parking tickets	Self-report	Possible shared variance due to the fines being issued for accidents
McBride, 1970	r?	-.365?	75	State record?	Mileage in one year	Self-report	High violation/accident drivers
McGuire, 1956a	Pearson	.44	134	State record	Moving violations	Army records	Contrasted groups added together Data from table 2 plus controls
McGuire, 1956c	Pearson	.03 .30	67 57	Self-report	Moving violations	Self-report	All drivers had had accidents in 'recent months'. Violations without accidents
McGuire, 1972	Pearson?	.24 .32	1,481 1,480	Self-report	Moving violations	Self-report	Some uncertainty about the time period, might have differed somewhat between subjects
Mesken, Lajunen and Summala, 2002	Pearson?	.56	1,160	Self-report	Fines (all)	Self-report	–
Owsley, Ball, Sloane, Roenker and Bruni, 1991	Pearson	.34	53	State record	Convictions for violations	State record	–
Peck and Kuan, 1983	r	.165	87,908	State record	Convictions	State record	Non-concurrent prediction
Peck, McBride and Coppin, 1971	r	.116 .115 .100	86,726 86,726 86,726	State record	Convictions	State record	Males. Each conviction could contain several violations

Table 7.3   Concluded

Study	Statistic	Effect	N	Accident source	Predictors	Violation source	Comments
Peck, McBride and Coppin, 1971	r	.072 .073 .069	61,280 61,280 61,280	State record	Convictions	State record	Females. Each conviction could contain several violations
Perry, 1986	Pearson	.46	54	Self-report	Tickets for violations	Self-report	–
Shope, Waller, Raghunathan and Patil, 2001	Pearson?	.280 (men) .206 (women)	4,403	State record	Serious offences	State record	Offences were alcohol-related, speeding>15 miles, reckless driving, vehicular homicide, other major offences, non-driving drug offences
Simon and Corbett, 1996	Pearson correlation?	.13	406	Self-report	'Offending score'.	Self-report	Dependent variable a composite of crashes, responsibility and exposure. Composite predictor from unlawful driving behaviours and ratings in terms of dangerousness
Smith and Heckert, 1998	Pearson?	.38	100	State record	Moving violations	State record	–
Smith and Kirkham, 1982	Kendall rank correlation	.194	113	Self-report	All violations	Self-report	–
Sobel and Underhill, 1976	Pearson	.22 (males) .11 (females)	283 213	State record Self-report	Undefined violations	Self-report	Accidents probably combined between sources
Stewart, 1957	chi2 Phi2	2.83 .010	275	State record	Speeding citations	State record	Phi2 calculated from table 4. Chi2 for all citations was somewhat lower, so the association with accidents should be similar
Wasielewski, 1984	Pearson?	.29	2,571	State record	Violations	State record	Car drivers

? denotes uncertain reporting.

Note: Shown are the association measures used, the effects, number of subjects used, source of the accident data, the category of violations used for correlation and the source of violation data. In California, where many of these studies have been undertaken, negligent operator points are issued for moving violations, that is not equipment or registration deficiencies (Harano and Peck, 1972).

excluded, who did find significant associations between various violation types and crashes. The statistical methods used in that study, however, make the results impossible to compare directly to those in Table 7.3.

The secondary correlation method yields results for violations versus crashes that are very interesting (see Table 7.4). Some studies (for example, Harrington, 1972; Wilson, Meckle, Wiggins and Cooper, 2006) show fair agreement between the patterns of associations with the predictors for accidents and violations, while others are weak, or even negative. Furthermore, the mean sizes of the predictor correlations were larger for violations as compared to accidents in most studies, in agreement with the basic idea of proxies being used because they have more variance and thus yield stronger associations.

Some peculiar results have also been found when collisions and violations have been used in parallel as outcome variables. Ferdun, Peck and Coppin (1967) reported that the effects on these variables for age/experience actually were in different directions. Similarly, Janke (1994) reported a decrease in violations but a possible increase in accidents when reviewing five consecutive evaluations of a driver improvement program for elderly drivers. The same kind of result was also reported by Struckman-Johnson, Lund, Williams and Osborne (1989), while Lund and Williams (1985) reviewed the literature about the Defensive Driving Course and concluded that it reduced violations but not accidents. The explanation for such deviant results would seem to be that it is possible to learn to avoid violating the law (for example, by being trained), but that this does not make you a better driver in terms of safety, but possibly even worse (Janke, 1994). Also, the strength of the association between crashes and violations would seem to decrease with the age of the driver (Daigneault, Joly and Frigon, 2002, see Table 7.3), which means that for elderly drivers, this safety proxy variable is even less useful.

In conclusion, it is difficult to evaluate whether violations can actually be used as a safety proxy at the individual level, because the evidence is very differing between studies. Also, the basic problem of whether there is shared variance between accidents and violations due to tickets being issued for crashes need to be resolved. However, this is a major undertaking in itself, as several authors would have to be questioned about this aspect of their method.

It might not be obvious why the shared variance problem is a problem at all. However, it should be remembered that the use of a predictor, or proxy variable, is really most often as a predictor of accidents in a future time, that is, as an indicator of what will probably happen. But when crash and citation data share variance due to a common origin, the association found will not be applicable to a real predictive situation. Instead, if the association should be applied logically, accidents and violations should both be applied as predictors of future collisions. Some indications exist that violations have some predictive power in excess of that of previous crashes (for example, Gebers and Peck, 2000; 2003), and that such a method would therefore be viable. However, it should be pointed out that what is meant here is not the adding of these variables into one, but a multiple regression approach, where each variable adds its own unique explanatory power.

Table 7.4 Some studies that have used citations and crashes as criteria in parallel

Study	Violation variable	Correlation between effects	N	Original N	Original statistic	Mean effect accidents	Mean effect violations	Accident source	Violation source	Comments
Arthur and Graziano, 1996	Moving violations	.93 (total accidents) .03 (at fault) .77 (not at fault) .49 (total accidents/year) .13 (at fault/year) .80 (not at fault)	7	250	r	.152 .198 .045 .096 .141 .015	.146 .109	Self-report	Self-report	–
Arthur et al., 2005	Moving violations	.772 .816 .568 .720	12	333	r	.065 (1) .150 (0)	.081 (1) .133 (0)	State record Self-report State record Self-report	State record Self-report Self-report State record	Data from table 3
Brown and Beardie, 1960	Not defined	.81	13	993	r?	.046	.054	State record	State record	Data from table 3
Burns and Wilde, 1995	Speeding violations All violations except speed Demerit points	.37 .43 .71	8	51	r?	.130	.317 .259 .154	State record	State record	Demerit points mean effect .154
Clement and Jonah, 1984	Number of convictions	.60 (men) .64 (women)	6	130 155	Correlation	.082 .074	.056 .069	Self-report	Self-report	–
Dahlen and White, 2006	Moving violations	.80 (minor) .78 (major)	11	312	Correlation	.092 (minor) .075 (major)	.123	Self-report	Self-report	–
Deffenbacher, Lynch, Oetting and Swaim, 2002	Moving violations	.16 (minor) -.28 (major)	7	290	r?	.023 (minor) .101 (major)	.076	Self-report	Self-report	Time period 3 months. Data from table 4
Deffenbacher, White and Lynch, 2004	Moving violations	.51 (men, minor) -.22 (men, major) .57 (women, minor) .53 (women, major)	14	218 218	r?	.081 (minor, men) .068 (minor, women) .102 (major, men) .073 (major, women)	.051 (men) .102 (women)	Self-report	Self-report	Data from table VI

Table 7.4   *Concluded*

Study	Number of violations/tickets	Secondary correlation	N (predictors)	N (subjects)	Statistic	Mean effect (1)	Mean effect (2)	Violation data	Accident data	Notes
Dula and Ballard, 2003	Number of tickets, 2 years	.87 (total number of crashes) .97 (ever cause crash)	8	119	r	.168 (total) .205 (ever)	.223	Self-report	Self-report	Data from table 3
Furnham and Saipe, 1993	Traffic convictions apart from alcohol related ones	.294	14	73	Correlation	.092	.223	Self-report	Self-report	Data from tables 1–3 and table 5
Harrington, 1972	All accidents/ convictions	.835 (males) .866 (females)	84 84	8,126 5,789	r	.055 .047	.142 .087	State record	State record	Data from table 15
Knee, Neighbors and Vietor, 2001	Traffic citations excluding parking tickets	.12	8	>99	r	.143	.139	Self-report	Self-report	Data from table 1
McBride, 1970	Violations	-.606	17	75	r	.135	.129	State record	State record	Extreme violation/accident sample
Owsley, Ball, Sloane, Roenker and Bruni, 1991	Convictions for violations	.29	17	53	r	.171	.132	State record	–	–
Schuster and Guilford, 1962	Moving violations	.61	38	100	r?	.129	.162	State record	State record	Data from table 1
Simons-Morton, Hartos, Leaf and Preusser, 2006	All accidents/moving violation	.27	15	2,260?	Odds ratio	1.24	1.93	Self-report	Self-report	Data from table IV
Sobel and Underhill, 1976	Not defined	-.12	6	496	r	.208	.207	State record/ Self-report	Self-report	Data from Table 3. One outlier. If removed, the correlation was -.95
Wasielewski, 1984	Violation points	.93	7	2,632	Correlation	.067	.120	State record	State record	–
Wilson, Meckle, Wiggins and Cooper, 2006	Speeding offences Other moving offences	.86 .87	6 6	32,120 26,830	Risk ratio	1.200	1.165 1.090	Insurance company	State record	Data from table 4

? denotes uncertain reporting.

Note: Shown are the type of violations included, the secondary correlations between effect sizes for violations/citations/convictions and accidents with different predictors, the N of the correlation (number of predictors in study), the number of subjects in study, type of original statistic, the mean effect for the original predictors versus accidents and violations, and the source of the violation data used. Unless otherwise noted, the accident and violation data are assumed to be for the same time periods. Means calculated as the square root of the mean of the squared effects.

The construction of predictors is also less restricted methodologically, as long as the dependent variable is a pure accident measure. What is here discussed as proxy variables often started out as predictors of accidents, and should indeed continue to be categorized as such.

Speed

> One of the most consistently powerful predictors of accident involvement in the literature is speeding ... McKenna and Horswill, 2006, p. 164.

Speed choice would, along with violations, seem to be the most popular proxy variable there is (for example, Lawshe, 1940; Newnam, Watson and Murray, 2004; McKenna and Horswill, 2006;[8] McKenna, Horswill and Alexander, 2006; Paris and Van den Broucke, 2008). However, the widespread use of this variable actually indicates a logical error, where evidence from one type of study has been accepted as evidence for something that it is not really applicable to (for example, Evans, 2004), as discussed under the heading of Alternative Logic. In reality, the evidence for speed choice as an accident predictor at the individual level is somewhat scarce (see Table 7.5), and the strength of the associations reported not very impressive. Here, it can be noted that speed is fairly equal in predictive power as compared to other predictors at the individual level, like personality (for a review, see Clarke and Robertson, 2005) and violations. In that sense, the above quote from McKenna and Horswill (2006) is correct. However, what these authors were arguing for was the use of speed as a proxy variable, and this is a different thing. Being relatively powerful as compared to other variables does not necessarily make speed a good proxy. Actually, it could be totally misleading. This is something we simply do not know today.

However, the results at the individual level need not necessarily be decisive, as both variables are usually somewhat unreliably estimated. What is needed is again a comparison between effects, in terms of a secondary correlation between the results for accidents and speed. Such studies are summarized in Table 7.6. Again, the results are few and very disparate, and no real conclusions can be drawn.

Turning from the common behaviour of speed to another often used proxy variable, we find that one of the reasons why it can be found in many studies is that it is rather generic; risky driving.

---

8   It is somewhat funny to note that this paper, which used speed and violations as outcome variables (but not accidents), was named 'Risk taking from the participant's perspective: The case of driving and *accident* risk' (italics added).

Table 7.5   Studies that have tested the association between traffic accident record and speed

Study	Statistic	Effect	N	Accident source	Predictors	Speed source	Comments
Arthur et al., 2001	Correlation	.22 / .05	394	Self-report / State record	Typical driving speed	Self-report / Self-report	–
Arthur et al., 2005	Correlation	.13 / .08	333	Self-report / State record	Typical driving speed	Self-report / Self-report	The same sample as in the 2001 study, but later criterion
Horswill and McKenna, 1999	Likelihood ratio test statistic	2.77 / 15.86	651	Self-report	Video speed choice / Questionnaire speed choice	Objective measurement / Self-report	All accidents. There were also analyses of speed-related crashes only
Karlaftis, Kotzampassakis and Kanellaidis, 2003	Probit coefficient	-.09	>17,000	Self-report	Fast driving	Self-report	Nothing reported concerning items or descriptive results
Quimby, Maycock, Palmer and Grayson, 1999	r	.20 / .25 / .15	113	Self-report	Mean speed / Assessment / Relative speed	Objective measurement / Self-report / Objective measurement	20 + 20 minutes of test-driving. Somewhat contrasted sample in terms of speed distribution
Wasielewski, 1984	Pearson?	.05	2,571	State record	Speed on a two-lane road	Objective measurement	The reporting was unclear on whether the value of speed used was based upon a single measurement or several
West, 1995	Spearman	.19	376	Self-report	DSQ speed subscale	Self-report	–
West, Elander and French, 1992 Study I	Point bi-serial	.22	711	Self-report	DSQ Speed	Self-report	–
West, Elander and French, 1992 Study II	Point bi-serial	.47 / .42 / .43 / .37 / .34	48	Self-report	'Preferred' speed I / Maximum speed I / 'Preferred' speed II / Maximum speed II / Self-reported speed	Objective measurement / Objective measurement / Objective measurement / Objective measurement / Self-report	Preferred speed was not defined. These data were also reported in West, French, Kemp and Elander, 1993
West, Elander and French, 1992 Study V	Pearson	.41	108	Self-report	DSQ Speed	Self-report	The same data as in West, Elander and French, 1993

Note: Shown the statistic used, the size of the effect, number of subjects used, the source of the accident data, the type of speed choice variable and the source of data for speed choice.

Table 7.6 **Correlations between correlations for speed and accidents versus different predictors in various studies**

Study	Correlation between effects	N	Original N	Mean effect accidents	Mean effect speed	Accident source	Speed source	Comments
Clement and Jonah, 1984	-.43 (men) .80 (women)	6	130 155	.082 .074	.122 .093	Self-report	Self-report	The mean size of the correlations did not take into account the sign of the original value
Wasielewski, 1984	.61	7	2,632	.067	.111	State record	Objective measure-ment	–
West, French, Kemp and Elander, 1993	.91	14	48	.244	.346	Self-report	Self-report	–

Note: Shown are the correlations between effect sizes, the N of the correlation (number of predictors in study), the number of subjects in the original study, the mean effect for the predictors versus accidents and speed, and the type of predictor used and the sources of data for the outcome variables.

Risky Driving

> ... a training which enhances perceived risks would respectively suppress risky behaviors, reducing accident risk and promoting safety driving. Rosenbloom, Shahar, Elharar and Danino, 2008, p. 699.

There are several unvalidated scales of 'risky driving behaviour' and similar concepts that have been used within traffic safety research (for example, Zimbardo, Keough and Boyd, 1998; Deffenbacher, Lynch, Oetting and Swaim, 2002; Fernandes, Job and Hatfield, 2007). Although usually poorly defined, they would usually seem to be meant to canvass all sorts of driving behaviour that the respondent (these are almost always self-report studies) and/or the researcher believe to be dangerous. This type of measure is therefore usually totally subjective, using items about speeding, tailgating etc, variables with very low validity as predictors to begin with, even if they were reliably measured.

Table 7.7 summarizes a number of studies that have related various self-reported driving behaviours to accidents. As with all other predictors of accidents, these are fairly weak. There is therefore no evidence at the individual level for the validity of risky driving scales as proxies for traffic safety. We therefore turn to the secondary correlation method (Table 7.8), and find that what slight evidence there is show that these scales are rather different from accidents as outcome variables.

As it has not been possible to find any research which validates the self-reporting of risky driving behaviour, apart from the DBQ, as discussed in Chapter 2, there would seem to be no scientific evidence whatsoever for the use of such scales as criteria of traffic safety at the individual level. But what if the risky driving is not self-reported, but observed?

Table 7.7   Correlations between various behaviour proxy safety variables and accidents in various studies

Study	Predictor variable	Accident variable	N	Effect
Adams-Guppy and Guppy, 2003	Lost concentration Driving while tired	All accidents	640	-.027 .062
Dahlen and Ragan, 2004	Risky driving	Minor accidents Major accidents	242	.25 .14
Dahlen and White, 2006	Lost concentration	Minor accidents Major accidents	312	.15 .04
Dahlen and White, 2006	Minor loss of vehicular control	Minor accidents Major accidents	312	.12 .03
Deffenbacher, Lynch, Oetting and Swaim, 2002	Risky behaviour	Minor accidents Major accidents	290	.08 -.11

Note: Shown are the variables used, category of accidents predicted, number of subjects used and size of correlation. All data self-reported. All statistics were correlations.

Table 7.8   Correlations between correlations for various behaviour variables and accidents versus predictors in various studies

Study	Behaviour variable	Correlation between effects	N	Original N	Mean effect accidents	Mean effect behaviour
Dahlen and White, 2006	Lost concentration	.79 (minor) .54 (major)	11	312	.092 .075	.142
Dahlen and White, 2006	Minor loss of vehicular control	.73 (minor) .52 (major)	11	312	.092 .075	.183
Deffenbacher, Lynch, Oetting and Swaim, 2002	Risky behaviour	-.64 (minor accidents) -.62 (major accidents)	7	290	.023 .101	.296

Note: Shown are the type of behaviour, the correlation between effect sizes, the N of the correlation (number of predictors in study), the number of subjects in the original study, the mean effect for the predictors versus accidents and behaviour. All data self-reported.

One outcome variable that might be subsumed under the heading of risky driving without being self-reported is that of a driving test, that is being assessed while driving by some sort of instructor, researcher or official, presumably with enough knowledge to make such an assessment. It should be remembered that what is at stake here is not really whether the assessor is able to notice whether the subject is driving too fast, overtaking on blind corners etc, but whether he/she can predict with any accuracy the future accident rate of the subject (this prediction is most often dichotomised as pass/fail). There is precious little evidence regarding the validity of such driver assessments (for example, Jonah, Dawson and Bragg, 1981; Groeger and Grande, 1996; Baughan and Sexton, 2001), which has yielded extremely low coefficients, but this has not stopped researchers from using it as a criterion variable.[9] Most interesting here is the conclusion drawn by Galski, Ehle and Bruno (1990) when their tests could not predict the outcome of a driving test: 'These findings raise serious doubts about the validity of perceptual and neuropsychological tests to assess the skills and abilities required for safe driving.' (p. 711). Here, 'perceptual and neuropsychological' should have been replaced by 'driving tests'.

Various Other Behaviour Proxy Measures

> ... the authors acknowledge that ... the criterion validity of the subtest behavior in traffic has to be examined using accident rates ... Arendasy, Hergovich, Sommer and Bognar, 2005, p. 318.

The previous sections have described the use of the most commonly used proxy variables, for which also some validity figures can be found. However, there are many other proxies to be found in the traffic safety research literature, which have hardly ever been tested against accidents, and therefore have largely unknown validities.[10] Here, a few of these will be discussed, along with some arguments and techniques of presentation used by researchers who employ such variables.

The use of unvalidated behaviour proxy variables is most apparent in the work of Deffenbacher and colleagues, where two such parameters have been employed in a fair number of studies, without ever being discussed as to their validity; loss of concentration and minor loss of vehicular control. In parallel, close calls, violations and minor and major accidents have been used as outcomes. The term used is 'crash-related outcomes' (or 'conditions'), so there is no uncertainty regarding

---

9   For example, Brookhuis and de Waard, 1993; Tarawneh, McCoy, Bishu and Ballard, 1993; Groeger and Grande, 1996; McDonald, Parker, Sutcliffe and Rabitt, 2000; Baldock, Mathias, McLean and Berndt, 2006.

10   As these proxies are often self-reported, it can also be questioned whether they have any validity as measurements of the constructs in themselves, but here only the more important matter of correlations with individual accident rates is considered.

the expected status of these parameters; they are certainly thought to be accident proxy variables. It can be noted that the term 'crash-related outcomes' is a bit strange, as two of the parameters are accident variables (minor and major), and thus not 'crash-related' at all, but the thing itself.

But turning back to the behaviour safety proxy variables used by the Deffenbacher group, it can be wondered how these are thought to relate to driving risk. Minor loss of vehicular control is probably similar to near misses, and although not strongly related to this variable, can probably be subsumed under the same kind of logic. Losing concentration is more difficult to understand as an outcome variable, in terms of safety. Does this include times when nothing at all happened? Does cruising along the highway and listening to the radio without really thinking about the driving make the grade? If there is any research that has shown that this is even remotely connected to accidents, these studies remain hidden.

The Deffenbacher research group has often reported that these 'crash-related items' do not form a reliable index[11] (Deffenbacher, Lynch, Deffenbacher and Oetting, 2001; Deffenbacher, Lynch, Oetting and Swaim, 2002; Deffenbacher, Petrilli, Lynch, Oetting and Swaim, 2003), that is, they were weakly inter-correlated. However, this did not prevent them from adding them into a scale in one study (Deffenbacher, Lynch, Filetti, Dahlen and Oetting, 2003). In general, the Deffenbacher research is, for many reasons, among the worst published; the use of self-reported, unvalidated proxy variables, sometimes for variable time periods, make their findings unusually prone to common method variance and generally unintelligible. As could be expected, the secondary correlations between the effects for their array of outcome variables are weak (see Table 7.8), and where such associations cannot be calculated, due to too small Ns, it can still be seen that they yield very different results. For example, in Deffenbacher, Lynch, Oetting and Yingling (2001), the correlations with their predictor were negative for (minor and major) accidents, but positive for all other.

Another driving behaviour variable of a most peculiar type was used by Carty, Stough and Gillespie (1998); change in driving style post accident (see Table 6.4 for secondary correlations with the effects of other outcome variables). What ever that variable is, it does not seem to have been used by anyone else. Although its status as a criterion variable could be questioned, it is an interesting research area that is very underdeveloped. As the stability of accident record calculations in Chapter 3 showed that people in general do not seem to change their driving after accidents, it could in principle be important to determine if someone does change, and why. Whether this reaction can be used as a safety outcome variable, is uncertain, however.

---

11   Cronbach alpha values of .4–.5.

Simulator Proxy Variables

> To the best of our knowledge, however, there is no published research documenting the correlation between simulators and crashes among older adults.
> Lee, Lee, Cameron and Li-Tsang, 2003, p. 454.

The use of driving simulators in driver research has a fairly long history. Examples of basic set-ups can be found as far back as in Häkkinen (1958). What is interesting from the present perspective is that a research tradition has evolved where driver behaviours in simulators are not used as predictors of traffic safety, but outcome variables, in effect replacing real-world accident data (see Table 7.10 and the review about marihuana and driving by Moskowitz, 1985). Even more interesting is that authors often do not even discuss the relevance of using such a twice-removed outcome parameter, or mention it only in passing (for example, Findley, Fabrizio, Knight, Norcross, LaForte and Suratt, 1989; McMillen, Smith and Wells-Parker, 1989). Again, face validity is all that is offered.

Even when the aim of a study is not explicitly about safety, variables thought to be important to it are used as measures, like speed (for example, Lee, Cameron and Lee, 2003), with the probable effect that the results are accepted as relevant for safety. This is an instance of the argumentational technique discussed in the logic section, where ostensibly the study is about behaviour, but conclusions are about safety.

It can therefore be asked what the associations between simulator driving variables and accidents are, that is, is there such a strong relationship with safety that simulator data can be used as a proxy? And if such association should exist, which ones of the myriad of different variables that can be measured should be used? It can be noted that the type of validity sought here is exclusively about the simulator as a replacement for crashes as safety criterion. This is rather different from how validity is normally tested for simulators, where the main interest has been in testing whether this type of driving is predictive of on-street performance (for example, Godley, Triggs and Fildes, 2002; Lee, Cameron and Lee, 2003), or if similar effects of predictors are found in the simulator as in other environments (Santos, Merat, Mouta, Brookhuis and de Waard, 2005). What evidence regarding the association between driving simulator variables and real-world crashes could be found is summarized in Table 7.9.

Many simulator studies refer to Edwards, Hahn and Fleischman (1977) for validity of their undertakings, and it is therefore of special interest to review this paper in detail. In the Edwards et al. study, taxi drivers were tested in two simulators, yielding sixteen different predictor variables. Self-reported and recorded accidents and violations were used as outcome variables, in a number of combinations, which was unfortunately not stated, but a conservative guess would be at least fifteen. The number of correlations run between variables must therefore have been more than two hundred. Out of these, two were significant (see Table 7.9). As two different sample sizes apparently were used, for self-reported and recorded variables, it is impossible to know exactly what correlation levels were present, but they were at least below .2.

The results in this study thus yield precious little evidence of any predictive power of simulator driving versus accidents or violations. It should also be noted that although the significant correlations, low as they were, might be forwarded as evidence, but not for simulator driving in general, but only for the variables included; 'correct braking' and total score. Although the latter overall score might be positively interpreted, the lack of strict definitions of how the variable was constructed makes this a precarious undertaking.

The Edwards et al. study is an example of a general problem apparent in the validation of simulator driving against accidents. There is any number of possible combinations of variables that can be measured and tested as crash predictors (or proxies), with slightly different mathematical definitions, and in various combinations with each other. The number of possible predictors therefore becomes very large, and the possibility of finding some spurious significant associations rather fair. If to this is added reactions of the drivers to special situations within the driving environment (as done by Deery and Fildes, 1999), the whole thing becomes totally unfathomable.

To further emphasise the problem of what simulator variables have actually been tested against accident record, it should be pointed out that the variables named in Table 7.9 are very poorly described in the studies testing them. This means that it is difficult to know exactly what it is that is predictive of accidents in a certain study, or if the simulator variable used as outcome in another is even similar to the one 'validated'. Exact mathematical descriptions of how physical variables are measured should be included in studies using such (see for example, af Wåhlberg, 2006a; 2007b).

Given the problems of a multitude of possible simulator variables and the scarcity of definitions, the rather common habit of researchers in this field to discuss validity in very general terms become even more problematic. Even when references are given, it is uncertain if the evidence they give is actually applicable to the study giving them (for example, Barkley, Murphy, O'Connell, Anderson and Connor, 2006).

In conclusion, driving simulator variables have very little backing as safety proxies for individual differences, and it is strange that such research can actually be funded, given the fairly large sums involved. This, of course, is not to say that simulators are not useful for other purposes, among which one of the most interesting today is driver training. For research on individual differences in driver behaviour, instrumented cars would seem to be a better option today, especially when the electronic revolution make them (comparatively) cheap, reliable and available.

Table 7.9 Studies that have tested the associations between various driving simulator variables and accident record

Study	Predictor variables	Accident source	Statistic	Effects	N	Comments
Cox, Taylor and Kovatchev, 1999	'Low risk/High risk*	Self-report	t	2.08	38	Effect calculated from raw data in paper. Contrasted groups. No definition of predictor
Deery and Fildes, 1999	Mean/std of speed Time to reach desired speed Maximum acceleration Time taken to initiate speed change	Self-report	h²	ns ns? ns? ns?	54	Cluster groups with different accident records were compared
Edwards, Hahn and Fleischman, 1977	Speed Braking Steering Signalling Crashes in various variants Total incorrect score	State record/Self-report?	Correlation	ns .25 (correct braking) ns ns ns .20	152/304?	It was not quite certain which source of accidents was used
Lee, Lee, Cameron and Li-Tsang, 2003	Rules compliance Traffic sign compliance Driving speed Use of indicator Road use obligation Decision and judgement Working memory Two simultaneous tasks Speed compliance Divided attention task	Self-report	Odds ratio	ns ns ns ns ns .39 0.55 ns 0.83 ns	129	–
Pizza, Contardi, Ferlisi, Mondini and Cirignotta, 2008	Crashes Time to first crash Divided attention Reaction time Sd of reaction time Sd of midline Mean speed Sd speed Speeding	Self-report	ANOVA	All ns	30	The group were all OSAS patients, and the variation within the sample may not have been enough for this calculation
Szlyk, Mahler, Seiple, Edward and Wilensky, 2005	Simulator crashes	Self-report	Correlation	-.19	35	–
Szlyk, Taglia, Paliga, Edward and Wilensky, 2002	Simulator crashes	Self-report	Correlation	.45	24	Glaucoma patients. Control group not reported but r<.43 for this group

? denotes uncertain reporting.

Note: Shown are predictor variables, the source of accident data, statistics used, effects and size of the samples.

Table 7.10   Studies that have used un-validated outcome variables in simulators

Study	Variables	Comments
Alm and Nilsson, 1994	Speed, lateral position, reaction time	No validation, no discussion
Barkley, Murphy, O'Connell, Anderson and Connor, 2006	Average speed, variability of speed, collisions, steering variability, total braking	No validation, some references
Barrett and Thornton, 1968	Emergency reactions (initial brake reaction and deceleration rate)	No validation, no discussion
Barrett, Thornton and Cabe, 1969	Initial brake reaction, deceleration rate ('emergency reactions')	No validation, no discussion
Findley, Fabrizio, Knight, Norcross, LaForte and Suratt, 1989	Steering, signalling, braking, accelerating and speeding errors	No validation. Errors not defined
Galski, Williams and Ehle, 2000	Acceleration errors, signalling errors, braking errors, steering distance, evasive action (braking, steering), threat recognition (braking, steering)	No validation?
Rogé, Pébayle, Campagne and Muzet, 2005	Speed, reaction time, time to collision	No validation

? denotes uncertain reporting.

Variables with Uncertain Status

> ... the justification for adopting proxy measures is often argued on the basis of logic, rather than empirical evidence. Clarke, 2006, p. 548.

Apart from the many proxy variables discussed, where some research can be found regarding their relations to accidents, there is also a group of parameters with a somewhat uncertain status. Although they are not always explicitly stated as outcome variables with an association to safety, they are nonetheless used as correlates to other variables in road user studies that seem to be about (individual differences in) safety. This includes, for example, seat belt use (Burns and Wilde, 1995), and 'hazardous situations' (Pöysti, Rajalin and Summala, 2005), a variable that might be very similar to near accidents. No definition was given in the study, and it is thus possible that the interpretation was left to the respondents.

Although it is often impossible to determine exactly what the assumptions of the authors are regarding the relations between their variables and accidents, a few of these instances will be discussed, to point out the problem. The defining features for the studies involved are that the variables are poorly defined, and no references given regarding validity.

Risk Perception/Accident Likelihood

> ... it is unlikely that accident involvement will capture all aspects of risky driving
> behavior, since only a few of the risky actions taken in traffic situations can be
> expected to lead to an accident. Ulleberg and Rundmo, 2002, pp. 235–236.

One possible quasi proxy variable that has surfaced in recent years is the questionnaire item 'accident likelihood'[12] (for example, McKenna and Horswill, 2006), which is probably roughly the same thing as 'risk perception' (Rosenbloom, Shahar, Elharar and Danino, 2008). This is an example where the stated relation of the variable would seem to be rather far removed from actual risk or safety. In fact, it is described as; 'Questionnaire measures of ... accident likelihood may be more useful in gaining an understanding of the strategic components that can influence behaviour. Such strategic factors might take the form of general attitudes surrounding ability that can in turn influence whether drivers expose themselves to hazardous driving scenarios.' (Farrand and McKenna, 2001, pp. 210–211). As should be apparent, the claimed association was several steps removed from anything resembling a measure of actual risk, and not even this very peripheral, hypothetical association was backed up by any evidence.

However, as the status of the accident likelihood variable is often not explicitly stated (for example, McEvoy, Stevenson and Woodward, 2006; Machin and Sankey, 2008), it is possible that it is intended as a predictor and not as an outcome variable. This would seem to be the case in McKenna and Horswill (2006), where a video speed test and the DBQ violation scale were instead used as 'risk-taking measures'.

In conclusion, there would seem to exist a number of questionnaire items/ scales that are not explicitly defined as to their intended use. This is a worrying development (because it would seem to be fairly new), and another indication of the lowered quality of traffic safety research. The next step would seem to be sheer randomness of scale construction and use.

Discussion

> I'm afraid that much research in traffic accident prevention is done with methods
> similar to this. Heath, 1962, p. 4.

Within traffic safety research, proxy variables of all types seem to be accepted, without this even being discussed, or evidence for their validity demanded or provided (for example, Evans, 2004). This means that studies can be found using criteria such as mobile phone use (Sullman and Baas, 2004) and driving test

---

12   The actual item is 'Compared to the average driver how likely do you think you are of being involved in a driving accident?' Farrand and McKenna, 2001, p. 204.

outcome (van Zomeren, Brouwer, Rothengatter and Snoek, 1988; Hunt, Morris, Edwards and Wilson, 1993; Bouillon, Mazer and Gelinas, 2006), neither of which has a strong association with traffic safety, in terms of individual differences. This does not necessarily mean that these studies are erroneous in terms of the associations found, but that any interpretation regarding individual differences in safety is very unreliable. Any variable that is associated with crashes is interesting and important, but only to the degree of the actual association.

One thing that becomes rather apparent when the arguments for proxy safety variables are studied is that many researchers seem to think that various replacements are somehow more valid as indicators of safety than accidents (as explicitly stated by Anstey, Wood, Lord and Walker, 2005). This may be due to a general distrust of all sources of accident data, but would also at times seem to go beyond that methodological objection. How this could be the case does not ever seem to have been stated with sufficient clarity to be understood.

The assumption underlying the accident proxy variables would seem to be that there is a rather strong association between crashes and these variables, and the latter can therefore replace the former as dependent variable. Exactly how strong this relationship should be to make the replacement acceptable does not seem ever to have been stated.

Looking into the actual data for the validity of proxies is therefore very much a matter of saying that the emperor is naked; what few fibres of evidence there are, actually covering the bodies of proxies, are easy to see through, once they are displayed together (its an ugly sight). The evidence tells us that not only are the associations between proxies and accidents weak, predictors also have different correlation patterns with these parameters (although this evidence is somewhat scarce). And this is the good part; the bad part is that many researchers use proxy variables where it is not even possible to find any data on their association with accidents. Actually, for some of them, it is hard to even understand how they are thought to possibly relate to safety at all. It must therefore be concluded that the use of proxies is a doubtful undertaking; they are not valid replacements for accidents themselves as outcome variables for individual differences in traffic safety.

The near-accident variable bears a close resemblance to the 'traffic conflicts' technique used by traffic technicians for safety evaluations of places. Interestingly, the same problem seems to be present as that found in the present data; what few validation studies have been undertaken have shown low figures. The difference would rather seem to be that for the traffic conflicts method, this has been pointed out (for example, Williams, 1981; Grayson and Hakkert, 1987), while in traffic psychology, the use of proxy variables has been unchallenged (not a single instance of a critical view against this practice has been found, or against any type of proxy).

That the (secondary) correlations between effects for proxies and accidents are rather weak is apparently a new finding, but the effects for the direct correlations have been there to see for decades. It can therefore be asked why proxies are used, because it cannot be argued that they add relevant information (but rather conflicting such) to the traffic safety literature. Traffic safety can only be one thing, and this variable must be best approximated by some crash variable. The use of a proxy variable in parallel to, not to mention as an amalgate with, accidents, must be regarded as illogical. The only explanation for its use would seem to be that it increases the probability of finding significant effects. If it is actually related to safety is apparently secondary for many researchers. It is simply a case of wanting to show results without having to grapple with the notoriously difficult crash variable, and have to admit failure in predicting it. Then again, it should be pointed out that it would be a wholly acceptable undertaking to relate predictors to proxies, if no claims were made about the latter's association with safety. But this is not how it usually works; instead, the researchers want to eat the cake and keep it too; they take the easy way out of their methodological problem by using a proxy that is simple to measure and will certainly yield some nice effects, but they also, directly or indirectly, claim that their proxy is indeed related to safety, and can therefore be seen as the real thing. Also, most of the variables studied would be fairly uninteresting if they were not thought to relate to safety.

There is also another reason for finding the use of proxies to be somewhat illogical. The main reason for their use is that accidents are weakly associated with all variables studied so far, including all that have been used as proxies. But this reason automatically invalidates this use, and if a really strong predictor of accidents was actually found, the need for studying other variables would lessen, which then would not need to be tested against anything. A good proxy would simply undermine its own usefulness.

It can also be pointed out that in some studies the use of violations/tickets as outcome would seem to be totally erroneous, because it is difficult to see how the kind of behaviour that leads to infringements of the law can result from the variables studied. Such a case is, for example, epilepsy (for example, Bener, Murdoch, Achan, Karama and Sztriha, 1996). Although seizure attacks may have the effect of temporarily loss of control and some type of infringement, this can hardly be such frequent events as to result in a violation record of any size. Such a choice of methodology would therefore seem to indicate again the carelessness with which most traffic safety researchers construct their driving safety criteria. However, one result has been found that is supportive of the use of citations as a proxy variable even in groups where this would seem to be wrong. Johnson and Keltner (1983) found that drivers with visual field loss in both eyes had twice the number of convictions as compared to controls. How such an effect might come about would seem to need an explanation.

What do the low correlations between accidents and proxies mean? Is it just a statistical and methodological problem that results in these weak associations? Or is there really a rather weak association between these phenomena? The latter explanation is not only worrying, but also rather intriguing, as it goes against all accepted gut feeling. How would it, for example, be possible that drivers could have close calls without accidents, and/or vice versa (one of these would probably be sufficient to explain the lack of association)? It could be speculated that drivers of the first category are very adept at reading the road, but use very small safety limits, that is they are conscious risk takers who get away with it. The second group would rather be the opposite. It would not take many such drivers to insert enough error into a calculation to distort the association for the other.

Anyway, it can be observed that, in similarity with how accident criteria are calculated, it can be suspected for simple reasons of probability, that most of the proxy safety variables used have low validity. There is just such an array of different variables used that not all of them can be right, unless they are rather heavily correlated, something that has not been shown.

How strongly should a parameter be associated with accidents to be considered useful as a traffic safety proxy in research? As has been shown, most direct associations are rather weak (less than ten per cent shared variance), and given only this correlation as a criterion, none really would do of the variables used so far. However, such weak associations can always be explained by the difficult statistical nature of accidents, as usually used within research. Therefore, the secondary correlation analyses undertaken become much more important, because they show whether the results come out with the same relative strength between predictors. If they did, the only difference between a good proxy and accidents would be stronger effects for the former, making it easier to detect the differences sought, just as intended. However, the question remain, what should be considered a good proxy, in terms of a secondary correlation between effects on various predictors, or in a direct correlation with accidents? The users of proxy variables need to define their position regarding such questions.

What variables can be used as proxies for accidents? As usual, it would seem that no really easy answer is forthcoming to such a question, but that a relative one may be given. Given the results in the present analyses, it would seem like the best proxy is state-recorded violations. However, as the present review of research was not exhaustive, and the studies used not intended for the purpose of establishing the relative validity of proxies, further work is certainly needed, and may yield interesting insights. For violations, the methodological problem of shared variance due to tickets being issued for crashes remain, and research in this area may change the current belief in this variable.

As noted, it could be argued that the weak association between a proxy and accidents is not really indicative of the true relationship, and that it is mainly due to measurement problems. There are two errors involved in this argument. First, it could be applied to any (lack of) association with accidents, regardless of everything else. If anyone wants to claim this, they therefore need to give some proof for it. Furthermore, even if it would be a measurement problem, this does not mean that the use of any proxy is better than accidents (or at least that it cannot be shown), or can yield any extra information. Whatever the real associations are, the only ones we can actually use are those we find empirically. This also means that it is usually useless to 'correct' an effect for measurement unreliability to get a 'truer' value (see for example, Clarke and Robertson, 2005). This 'corrected' value for an association will be a mathematical abstraction that does not make the prediction of actual accident involvement, for example, any more powerful. The only possible use for such a mathematical treatment would seem to be when someone finally undertakes a comprehensive meta-analysis of all known accident predictors and test whether their combined power may actually explain all variation in accident rates. Unfortunately, it would seem possible that the result of such a calculation would exceed 100 per cent explained variance, as many predictor variables intercorrelate. Also, a correction for the mean number of accidents in the samples could yield a kind of 'true' value for how strong the association would be under 'ideal' circumstances.

The uncritical assumptions about proxy safety variables sometimes lead to somewhat humorous results. Consider the claim by a researcher at a conference (Mann and Sullman, 2008), that previous unsafe pedestrian behaviour predicted later unsafe driving behaviour (both variables self-reported, of course). The type of 'unsafe' behaviour used as a predictor was crossing the road outside of zebra crossings. However, the available evidence (what little there is) rather seems to indicate that it is safer to cross anywhere but at the marked crossings (Herms, 1972). So, in the Mann and Sullman study, it would seem like safe behaviour predicted unsafe behaviour (this was also the case in Elliott and Baughan, 2004). Given that people like to appear to behave congruently, it would seem like they least reported what they considered to be safe in both situations, but were wrong about the crossings. This example should make it clear that the use of proxy parameters is not always such a good solution to the safety variable problem. As shown in this chapter, one has to know the actual association of a proxy with safety, or the research will be unintelligible.

The line between logically acceptable and unacceptable use of proxies can be hard to discern, however. In principle, it is correct when the research undertaken is not about individual differences (and there is evidence of validity). For example, Simons-Morton, Lerner and Singer (2005) studied the influence of passengers on various driving behaviours of young drivers. Their overall term was 'risky driving', clearly indicating that the study was about safety, using proxy variables. This was a valid method, because the study was not about individual differences, but about a within-individuals effect, measured at group level. Actually, what the authors probably wanted to measure was the mean effect on driver behaviour of the presence of passengers, while in reality they measured the difference between drivers with and without passengers. It can be expected that the difference between these methods was slight, but it is possible that riskier drivers tend to carry more passengers (an individual differences question).

Whatever the reasons for the results presented in this chapter, it must, for the time being, be concluded that the use of accident proxy variables should not be encouraged. Although it is a very tempting alternative, which almost always yields some significant differences, it is not scientifically proven to be valid, for any variable.

Chapter 8
Cases Studies

The previous chapters have presented hypotheses, arguments and data regarding a number of traffic accident research problems, often with short examples from various papers, mostly regarding peculiar methods used by various researchers. Here, a further few such examples will be given, but in more detail. They can be seen as case studies of how things have gone wrong within traffic safety research, and how the data that have been presented are essentially useless, or, at times, illuminating in ways that the original authors did not intend.

The Unproven Validity of Self-Reports

> ... self-report ... measures cannot be very helpful as independent variables where inferences are to be made about their effects on other self-report variables. Spector, 1994, p. 387.

The overall view of traffic researchers who use questionnaires would seem to be fairly well expressed by Hatakka, Keskinen, Katila and Laapotti (1997) in a paper enthusiastically defending the use of self-reports. Most of the arguments discussed in Chapter 2 can be found in this piece. However, they also made a number of claims and arguments that need some closer scrutiny. First, it should be noted that only arguments and indirect evidence was put forward, but no validation at all in the sense used in this book; checking the results of self-reports against more reliable sources. Actually, these authors did not refer to this possibility or type of research at all. For a group of researchers who claim 'In order to find the pitfalls and possibilities of research with self-evaluations, the methodological aspects should be carefully evaluated.' (p. 297), this is somewhat remarkable. Their style of argumentation was also strange, as they made good use of the old rhetorical trick of building a straw man. Thus, they said '... the whole idea that self-reports and behaviour would *not* be connected to each other is at least peculiar.' (p. 297, italics in original). However, they gave no reference to anyone who has ever made such a claim, and no one ever has, of course. Re-stating the opponents' claim in an extreme form is just a way of easily being able to brush it aside.

Instead, data from two waves of questionnaires on risky driving, skill and other fairly standard variables was reported, and correlated with (self-reported) collisions and violations (citations). The conclusion was that '*It is possible to predict drivers' violations and accidents ...*' (p. 301, italics in original). In the present work, such results would rather be interpreted as 'It is possible to predict drivers' self-reported violations and accidents ... with other self-reported items', which is no great feat, because common method variance would supply the basic association. And no italics would be used.

What is interesting about the results of Hatakka et al. is instead what they did not analyse at all; the pattern of data surrounding the only objectively gathered variable they had included; recorded violations in one of the samples. Here it should be noted that some of the details are guesswork, as the description of the method was sketchy, to say the least. For one thing, it was not stated if the two Violation variables had the same exact definition.

The study consisted of two samples of young drivers who completed a questionnaire, containing questions on driving while being in an improper state (presumably drunk), driving with high speed, violations and accidents. The aim of the study was then to use other items to predict these four variables. Analyses were undertaken separately for the two samples, yielding a table (number 1) of correlations between eight dependent and ten independent variables. However, for some reason, the correlations were not reported in numbers, but converted into three categories; non-significant, negative and positive. Although this method has as its result a loss of information, it is still possible to check how homogeneously the predicted variables relate to the predictors by replacing the three categories with the numbers zero, minus one and one. Thereafter (regarding each column as a variable), (secondary) correlations can be run between all eight dependent variables. The mean correlation of effects with all the other seven variables ran from .43 to .71 for each variable. The lowest value was found for Violations in sample 1.

Now, the interesting thing is that this variable was not self-reported. The analysis undertaken here therefore shows that the self-reported variables tended to go together, while the one objective parameter was different from the others. As expected, self-reported items tended to be more similar in their results. Actually, the next lowest mean correlation was 50 per cent higher than the one for objective data.

The analysis here is fairly crude, and should of course be undertaken on the original data, but it is strange that authors writing about the validity of self-reports would not try to analyse their data in some way that would yield a figure on how subjective and objective variables relates. Instead, they reported in detail about the presumably best results (from their point of view), the prediction of self-reported Violations in sample 2, which had three positive correlations in their Table 1, but disregarded that the same (but objectively gathered) variable had three negative and one positive correlation with the same predictors.

From such results, it was concluded that '... *the connections between self-evaluations and behavioural variables are strong.*' (p. 302, italics in original). But as the only objective variable had the weakest association with other variables, 'behavioural variables' presumably meant self-reported ones, and, as pointed out before, finding such connections is not very impressive. Actually, it could be expected that people would tend to answer in a congruent way even about totally fictional behaviours and states of mind, given that they thought the questions were about the same topic.

Differential Accident Involvement Versus Accident Proneness

As noted, the concept of accident proneness has fallen into disrepute with most researchers who have even paid attention to the issue for the last decades of the twentieth century. Instead, it has been suggested by McKenna (1983) that accident proneness should be replaced by 'differential accident involvement', which is (should be) '... an area of study not a set theory.' (p. 69). Although this alternative has not really caught on, it is at times referred to, and it is therefore of some interest to contrast it against the accident proneness thinking.

McKenna defined differential accident involvement as the issue of whether it is possible to predict crash involvement from psychological tests, whereas accident proneness was seen as statistical modelling, that is, distribution fitting. Furthermore, the first does not assume that accident involvement is a stable phenomenon, or any specific shape of distributions, or if accident involvement is a general or specific factor (which probably mean whether it is stable over environments). From these definitions, McKenna concluded that '... the study of differential accident involvement has the advantage that it makes fewer assumptions and is less open to the criticisms which have plagued the concept of accident proneness.' (p. 69). That differential accident involvement have fewer features than accident proneness is apparently true. Actually, it would seem to be a purely descriptive term, without any of the theoretical features of accident proneness. This, however, is not an advantage at all, but a scientific disadvantage, because the empirical research on accidents is reduced to random testing of effects. With the exception of the statistical distributions, the concept of accident proneness as some sort of innate characteristic of the individual does lead to a number of testable predictions, which is the hallmark of science (Popper, 1968).

It would seem as though McKenna also recognized this shortcoming of the new concept, as he stated that 'A theoretical understanding of differential accident involvement would lead to a knowledge of the psychological processes underlying human error.' (p. 69). Why a theoretical understanding of this idea, but not accident proneness, would have such effects was not stated. Furthermore, it is difficult to understand how a descriptive term could lead to any theoretical advances. It is therefore not the least puzzling that the concept of differential accident involvement has not been theoretically developed since the publication of the 1983 paper. It is simply very difficult to envision any basic theoretical ideas about individual differences in accident involvement that can explain this but does not need to take into account the ideas of innate, (at least semi-) stable characteristics of individuals that predispose them to have different numbers of mishaps. As noted, the whole idea of being able to predict accident involvement by fairly stable characteristics, such as personality, assumes stability of accident record.

The conclusions drawn here about accident proneness versus differential accident liability are thus the opposite of those of McKenna. The first is seen as a theoretical statement that predicts testable outcomes, while the latter has no useful scientific features. Instead, it confuses the issue, due to the claim that it can replace accident proneness as the basis for research in individual differences within this field.

However, it can be agreed with McKenna that the accident proneness concept has been used in very confusing, contradictory and logically circular ways. This, however, should not be a basis for disregarding it, as no real proof has been forwarded that the basic assumptions are wrong. Instead, this points to the need of theoretical development; the accident proneness idea need to be coherently defined and explored in terms of what could constitute its basic assumptions and what their testable consequences are. This is also a challenge to the concept of differential accident liability and its adherents, if any; is it even possible to develop this competitor into something resembling a scientific theory?

Constructing a Multitude of Safety Criteria

The use of several safety outcome variables in parallel is fairly common in traffic safety research. However, it should be stressed again that there can really only be one safety, and that the effect sizes for different dependent variables should therefore be in the same relative association with the predictors. If not, at least one of the dependent variables is a bad conceptualization and/or measurement. The record holding study regarding number of safety variables is probably Carty, Stough and Gillespie (1998), where eleven different dependents were used. As a whole battery of predictors were also included in the study, and some of these were comprehensively reported on (in terms of association values), the secondary correlation method can be used, and yields as many as fifty-five correlations between the effect patterns of the dependent variables (see Table 6.4). It can be seen from these correlations and a factor analysis that 'All accidents', 'accidents 2' and 'accidents 3', 'change in driving style' and 'dangerous driving convictions' seem to form a syndrome of fairly closely related dependents, while the others are solitary constructs.

Turning to the text of Carty et al., the reason for choosing it as an example was the unconcerned use of a whole host of accident variables. The following errors can be noted in conjunction with them:

1. It is uncertain what sources the data was gathered from, although it was probably all self-report.
2. The description of the safety variables is confusing, to an unprecedented degree. It is, for example, not possible to know what the labels in table 1 refer to without a very close reading of the text, and some guesswork

3. The time period for reporting is probably variable (time for employment ranged between 2 months and 13.5 years)
4. No definitions for 'preventable' or 'responsible' were given.
5. It was not explained why the variable 'change of driving style post accident' was included as a criterion, or how it was conceptualized.
6. A mixing of different groups of drivers in the sample (interstate and local heavy truck drivers with forklift chauffeurs).
7. Uncertainty whether the variable 'All accidents' included private driving incidents.
8. Only scattered reporting of descriptive values for most of the psychological inventories used.

The study of Carty et al., can therefore be used as a sample card of the various mistakes described in the previous chapters. Furthermore, to this can be added the heavy emphasis on discussing psychological mechanisms while totally disregarding the outcome variables. Also, the authors seem not to have reacted upon the very different patterns of associations for their different outcome variables, but picked out the significant values for discussion, regardless of what variable it was for.

If such 'research' as this is accepted as valid in any way, it is indeed easy to find significant associations and contribute to the literature. All one has to do is to use a dozen different 'accident criteria' and whatever number of questionnaires the poor subjects agree to fill out. Effects are bound to follow. This industrial production of artefactual results is very evident in the Carty et al. study, but is unfortunately very common within traffic safety research, although the errors are usually not as strongly clustered as in the presently criticised report.

Errors in Calculation of the Crash Variable

Even fairly simple logical/mathematical errors occur in traffic safety research, when the dependent variable is to be calculated, even when the necessary information for countering them is present, as is evident when the authors themselves have included it in their report. Thus, Rolls, Hall, Ingham and McDonald (1991) required subjects to report their total number of accidents and divided this total by the number of years they had held a driving license. This method resulted in three different errors; first, the authors had themselves stated that about 10 per cent of drivers reported illegal driving before being licensed. This means that those who had driven illegally probably included incidents from that time in their answers (the accident item did not specify whether it was licensed driving that was canvassed), but their total number of crashes would still only be divided by the years of licensed driving, thus inflating the accident rate for these people.

Second; apparently, the number of accidents per year reported was for a driver's whole career of driving. But this would give a very strange estimate of the safety of a middle-aged driver, as the accident record of the youthful years would still burden the calculation. Therefore, the results would not seem to be pure estimates of the older drivers' safety, but a serious under-estimation of it. Fortunately, the third error to some degree countered this bias, as the memory effects (forgetting and amalgation of events) would make older drivers report less accidents from their younger years.

This analysis was made possible by the fact that these authors actually described their method in some detail. In many cases, descriptions are instead fairly obscure, and it can only be suspected what has actually been undertaken, and what effects this might have had on the results.

A Very Alternative Criterion Variable and Method for Validation

> Chaloupka and Risser (1995) investigated the correlation between police-registered accidents for 51 road sections along a standardized route in Vienna and behavioural data obtained with the Vienna Driving Test. The authors reported significant correlation coefficients from $r=.38$ to $.48$ between observational variables obtained in the Vienna Driving Test and police-recorded accidents. Overall, these results argue for the validity of standardised driving tests. Sommer, Herle, Häusler, Risser, Schützhofer and Chaloupka (2008), pp. 363–364.

One instance of a proxy traffic safety variable is the Vienna Driving Test (VDT; Chaloupka and Risser, 1995), which is used for assessment of drivers in Austria. Although this would seem to be a fairly run-of-the-mill driving observation test, there is one thing that sets it apart from other tests. Instead of simply ignoring the matter of predictive validity (can the observations actually predict crash involvement?), as is usually the case for these tests, the originators have proposed an entirely new method for validation, apparently inspired by the traffic conflicts technique. From their findings, they claim that the VDT has good validity. Here, the method used to arrive at this conclusion will be scrutinized, but also the supplemental argumentation regarding the use of proxy variables. The main paper to be discussed here regarding the VDT and its validation is Sommer, Herle, Häusler, Risser, Schützhofer and Chaloupka (2008), due to the arguments and claims made in that publication. The main problem discussed is how to measure 'fitness to drive'.

The argumentation was:

1. All sources of accident data are unreliable; therefore, proxy variables should be used as criteria instead.
2. Driving tests can take into account near accidents, and are less prone to impression management and self-deception; therefore they can be used as proxies.
3. The VDT has been validated; therefore it can be used as a measure of individual differences in traffic safety.

The first argument used is a sort of development of the previously discussed crash record unreliability idea. As noted, showing that one source is unreliable does not have any impact on the alternative. In Sommer et al., the low correlations with self-reported accidents seem to be forwarded as some sort of evidence of the low quality of records. Thereafter, self-reports too are criticized as unreliable. The authors therefore seem to end up in a logical contradiction, where both correlated sources are seen as being of low quality, each referring to the other. But the only argument forwarded for the unreliability of records was that of a low correlation with the unreliable source of self-reports.

The second argument was two-fold, claiming positive features of driving tests. Regarding near accidents, it should be evident from Chapter 7 that there is no evidence that these should have any validity as a proxy variable regarding individual differences in traffic safety. It can here be suspected that Sommer et al. have made the logical mistake of transferring group level results (traffic conflicts research) to individual differences. Whether this is really so is hard to determine, because the references given are not cited in support of any real statement regarding the validity of near accidents.

Turning to impression management and self-deception, it should be noted that the statement in the summary above is verbatim; what the authors say is that driving tests are *less* prone to these problems, without stating what they are comparing them with. Also, no comparative data was shown, but a reference given for a review (Hjälmdahl and Várhelyi, 2004). In this review, it was stated that some authors had found differences in driving behaviour between driving tests and normal driving. There are precious few such studies, though, and it is uncertain exactly what variables should be measured, although acceleration behaviour would seem to be strongly influenced (af Wåhlberg and Melin, 2008). The claim about less effects of social desirability in driving tests was therefore totally unsubstantiated.

The claim regarding validation of the VDT in Sommer et al. gives a reference to Chaloupka and Risser (1995), and it is said that these authors circumvented (sic!) several of the methodological shortcomings of other studies on the criterion validity of standardized driving tests (what these were was not stated). This great leap forward was achieved by correlating the observations of the driving assessors at certain points along the standardized route, with the number of police-recorded crashes in these places. As should be evident from the section about intra- versus inter-individual effects in Chapter 7, this is using an (aggregated) within-driver effect to argue for the validity of the measurement method for between-drivers safety. This means that not only were the correlations found (.38–.48) much stronger than what would be the results for individual data, but also that the effects were unrelated to how the method can be used to identify the risky driving behaviour of individuals. What was measured was not any individual risk level, but the risk level of the accident points, and the study actually showed that driving assessors (as a group) can identify dangerous road sections with some accuracy. This is an interesting result, but it has nothing to do with the assessment of individual drivers. This method is actually a sort of traffic conflicts method, where the goal is to validate these against accidents at certain points of the road. This was reported, for example, by Risk and Shaoul (1982), but not as a method for validating an individual differences measure.

Turning back to the claims made by Sommer et al., the quote in the beginning of this section was all that was said about the method of validation used by Chaloupka and Risser (1995). For the unwary reader, the information given would seem to point to a correlation between the behaviour and accident record of individuals (although the mentioning of 51 road sections would then be anomalous). Given such a method, the obtained correlations would have been impressive. Given the real method, they are not even relevant for the claim made.

Finally, it can be noted that one of the arguments presented by Sommer et al. for the use of proxy safety variables was the low reliability of accidents. It is therefore somewhat contradictory that, in the end, recorded accidents were used as criterion for the validity of the driving assessment variable. So, despite their conclusion that 'These findings constitute a challenge for the claim that accident rates can be regarded as a highly reliable and valid 'golden standard' in measuring respondents' actual driving fitness.'[1] (p. 363), these authors still refer to accidents as the criterion in their validation.

Then again, this quote may also be interpreted as not being about the unreliability of data gathering and record keeping, but about whether accidents are actually a reflection of 'risky driving'. If so, this again highlights the problem of defining 'risky driving' without reference to accidents. If we did not have accidents, whatever 'risky driving' was undertaken would simply not be interesting.

Summing up, the arguments and research concerning the VDT as a proxy for individual traffic safety are erroneous, logically faulty and misleading. It can be suspected that the VDT has no validity what so ever regarding its main intended use; identifying dangerous drivers. Blaming bad records for the inability of a method to predict anything useful does not make it any more valid, but as with so many methods, this argumentational technique has apparently been used with some success.

---

1   It should be noted that no reference was given for the claim that a claim had been made that accident rates are highly reliable and valid. In fact, this is an instance of the rhetorical device of phrasing the opposing view in a more extreme form, which is therefore easy to dismiss. That is the reason why both reliability and standard are put into the same sentence. Although some researchers would probably want to use accidents as a golden standard, hardly anyone would claim that the available accident data is highly reliable.

Afterword

Unsupported conclusions do not become valid because they stem from the only information readily available or one wishes to be sympathetic to a researcher who finds it difficult to do adequate work. McGuire and Kersh, 1969, p. 16.

After wading through the fairly vast literature on individual differences in traffic safety, there is a need to summarize, conclude, discuss and speculate about what it all means. The many discrepancies found between research practice and actual evidence are disconcerting, and lead to a general doubt regarding the 'knowledge' generated within (individual differences) traffic safety research, a doubt that tends to spread to other areas, both academic and applied.

The present chapter will therefore repeat the main findings and conclusions, and discuss some other observations. Finally, the sceptical view presented in this book will be applied to certain traffic safety topics, to challenge the standard views.

Summing Up

Many traffic accident prevention measures are based on opinions, beliefs, suppositions, hunches, and 'common sense', despite the fact that valid evidence often is available. Heath, 1962, p. 4.

The results of this book, summarized chapter by chapter, can be described in the following manner. Chapter 1 discussed the various possibilities of categorizing traffic accidents to achieve optimal power. Although almost no research seems to have been undertaken with the specific aim of achieving this goal, it was concluded that culpability is the most promising feature to take into account. However, theoretical development of the relation between various predictors and specific categories of collisions is needed, and might yield interesting benefits.

Chapter 2 reviewed the use, abuse, evidence and arguments regarding self-report data in traffic safety research. The commonplace occurrence of this methodological deficiency suggests there may be a severe problem. The available evidence shows that all self-reports of driving behaviour are not only unreliable, but probably prone to common method variance effects, something which has been largely ignored in the traffic safety research community, despite such problems being well known in other, similar, subject areas.

Chapter 3 studied the issue of stability of accident record over time, and a meta-analysis of available evidence showed that the previous conclusions regarding this problem have been mistaken. Accident record would seem to be exceedingly stable over time. The problem is really that it is difficult to measure, due to low variance and unreliable sources of data.

Chapter 4 analysed the matter of culpability for crash involvement, introduced in the first chapter, regarding how it is conceived and used by researchers and what the consequences are for the results. It was found that the issue is seldom discussed, and that the percentages of culpable accident involvement in different samples show that there is substantial disagreement about what constitutes responsibility for a crash. Some results of the meta-analyses undertaken in this chapter indicated that results of accident prediction studies are indeed influenced by how culpability is defined, in line with the suggestions put forward in Chapter 1.

Chapter 5 investigated the commonly accepted but very peculiar 'fact' of a curvilinear association between exposure and traffic accidents. To some degree, this result was found to be a kind of urban myth, because there is little evidence that support it. Furthermore, several of the results were probably influenced by self-report biases, which create a part of the curvilinearity in the studies that use such data. Finally, it was suggested that the remaining curvilinearity is to some part due to bad drivers driving less, while the opposite effect, claimed by some researchers to explain the increase in accident rate of older drivers, is probably exceedingly small. However, most importantly, no one has ever shown that such an effect (less driving causes lowered proficiency) exists. It is instead an illogical inference from the supposed curvilinearity of the accident-exposure association.

Chapter 6 tried to sort the bewildering array of safety criteria used in individual differences studies into categories, and analysing what the probable effects of the various methods are. As no comparative research was available, no conclusions could be drawn about what could constitute the best methodology for constructing an accident variable as outcome measure in a study of individual differences in traffic safety. However, it can be assumed from the results of Chapter 3 that the length of the time period used should be similar to the time over which the predictors are fairly stable. Practically, this means that most of the studies published on various topics should at least double the time period used.

Chapter 7 was a similar undertaking to the preceding one, looking at the many alternative outcome variables used in traffic safety research. As with accident criteria, the supporting evidence and logic is scarce, and it can be concluded that the use of non-accident dependent variables is an unscientific undertaking, as most of them have not been validated, and what few have some backing have actually been shown to be very weakly related to safety in terms of accidents. The logic of proxy safety variables is also deficient, and their use in individual differences studies should therefore be omitted until positive evidence of their validity is shown.

Themes and Conclusions

> After thoroughly reviewing these studies, those reporting favorable as well as unfavorable results, I have no confidence in any of them. Heath, 1962, p. 4.

The results of the present analyses are very different from all previous reviews and meta-analyses of individual differences in traffic safety, because they have not taken into account the many methodological deficiencies discussed here.[1] This means that the conclusions of those reviews are considered as not based on sound research, as they all include many studies that have used methodology that is not only unreliable, but also potentially misleading.

The main theme of this book has been the many methodological problems in traffic safety research, and the low standards of research in this area today, apparently due to the unfounded assumptions about various methods made by many researchers. Also, it has been implicitly argued and shown that there has actually been a lot of interesting and potentially fruitful traffic safety research undertaken through the decades, research which is now more or less forgotten.

Another theme, which can be seen as parallel to the assumptions idea, is that traffic safety researchers have taken very little interest in the accident variable. For some reason, a majority would seem to have the view that it is an uncomplicated parameter, apart from its irritating habit of not delivering the significant results wished for. This is usually blamed on the statistical characteristics of crashes (for example, Lajunen, Corry, Summala and Hartley, 1998), that is, it is claimed that the reason for their failure lies with this problem. To accept that the methodology of most accident studies could be strongly improved does not seem to be a popular alternative, and even less so that the hypothesis tested might actually be wrong.

The specific problems discussed in this book are strongly intertwined, and not only because they all seem to be caused by a disinterest in accidents, or because many researchers commit several of these errors simultaneously. Instead, they share features and inter-relate. Thus, the problem of determining culpability is aggravated by the use of self-reports, the lack of knowledge about the stability of accident record leads to the use of short time periods for criteria, which causes weak effects that probably prompt the use of alternative outcome variables, which delivers statistical testing significances despite the lack of rigour in constructing the variable, thus further boosting the general negligence regarding accidents.

One of the reasons for writing this book was the argumentational technique that is commonly used when writing papers and arguing for the methodology used; selective referencing. Very often, it is said something along the lines of 'crashes have repeatedly been found to be a strong predictor of traffic accidents...', and then one or two references are given in support of this claim. These may perfectly

---

1   For example, Adams, 1970; Lester, 1991; Findley, Levinson and Bonnie, 1992; Connor, Whitlock, Norton and Jackson, 2001; Anstey, Wood, Lord and Walker, 2005; Clarke and Robertson, 2005; Jerome, Habinski and Segal, 2006; Barkley and Cox, 2007.

valid and supportive of the claim, yet the claim is wrong, because there are a dozen other papers where this effect has not been found. If confronted with this, the authors can (possibly truthfully) say; 'oh, we did not know that'. But when large numbers of studies on various methodological subjects have been summarized, as here, this strategy becomes more difficult. Claiming ignorance about the various methodological problems involved in individual differences in traffic safety research will now become akin to saying that the earth is flat. Not that the researchers criticized in this book will want to read it, or accept the conclusions, but sooner or later they will have to acknowledge its existence and start arguing against it.

One problem that might be noticed in the present text is that researchers sometimes have an economical interest in the thing they are researching. Currently, this would seem to mainly be the case for methods for determining driving risk (for example, Chaloupka and Risser, 1995; Sommer, Herle, Häusler, Risser, Schützhofer and Chaloupka, 2008; Darby, Murray and Raeside, 2009), where there is a fair market for various companies selling tests of unproven worth. Maybe it is time for traffic research journals to ask for statements about conflicts of interest. As it is, it is hard to know whether any specific author is benefiting from positive results being published.

One practical conclusion that can be drawn from the accident stability analyses is that it might indeed be warranted to exclude drivers with many (culpable) crashes from driving, as this is a very strong indicator of future mishaps. This has been denied by some researchers (for example, Summala, 1988), but the method as such seem only to have been tested by Rawson (1944), so there is no real evidence regarding this problem. Actually, there seems to be little research that has compared the predictive power of crashes versus other variables. Maybe the whole undertaking of trying to predict accident involvement from various other variables is misdirected, because the strongest predictor is previous accident involvement, if this can be established for a long period of time?

Given the arguments and data presented in this book, especially regarding self-reports and common method variance, it could be asked; what evidence is needed to convince hard-core self-report researchers that their method is worthy of suspicion? This question is not rhetorical, because it can be suspected that everything presented here will be dismissed with arguments about 'not enough data', 'too strong conclusions', and that lovely, totally unverifiable 'there is information in self-reports that you cannot get in any other way'. In short, there is probably nothing that will ever be accepted as evidence against self-reports by the hard-liners, no matter how many studies are made that show effects of common method variance and general unreliability of self-reports. However, this very unscientific stance is, of course, something that they will never willingly concede to. Therefore, the question is very important; what will you consider as evidence that self-reports are so unreliable and misleading that they should not be used (in conjunction with self-reported criteria)? This question is also rather funny to ask unsuspecting victims, because they do not have a ready answer, having never even entertained the idea that there could be something wrong.

As should be apparent from the various chapters of this book and the problems discussed, the arguments forwarded and the data presented, traffic safety research (especially traffic psychology) is in dire straits. Most studies would seem to use methods which can be criticized for various shortcomings, and although this could be said to be true for any discipline, the difference lies in that the problems discussed here do have remedies. Also, the conclusions drawn by many researchers simply do not match the restrictions imposed by their methodology.

However, the main conclusion from the analyses undertaken here is that there is a lack of research into methodological questions, which, up to a point, explains the general ignorance of this subject. We therefore turn to what kind of research is needed to improve the methodology of traffic safety research.

Badly Needed Research

How to Categorize Crashes

It is difficult to envision how research into accident categories should be undertaken, apart from tests of differentially constructed accident variables as outcome measures. In such studies, the theoretical basis for the construction of a variable should also predict which predictors should have the strongest associations with it.

Apart from this basic method, however, little is known about how to move forward in this area. There seems to be little, if any, theory to build upon, and common sense approaches will probably have to do for some time still. It can be noted that no existing theory of traffic accident causation seems to have mentioned that any specific category of crashes would be more predictable than any other (for example, Wilde, 1982; 1985; 1988; Summala, 1988; Fuller, 2005), apart from af Wåhlberg (2008b).

Common Method Variance and Questionnaire Validity

Research into common method variance effects is almost unknown in traffic safety. As a fair percentage of all studies published can be suspected of having this problem in some form, there is indeed a great need to study this problem.

Of course, the simple solution of not using self-reported outcome variables exists, which circumvents the problem, but somehow, it is difficult to believe that the researchers that are fond of self-reports will actually let go of this effortless method. It is equally hard to believe that they will undertake any CMV research, but, as usual, other researchers will have to step in and do the work that will probably show a great many deficiencies in other people's investigations.

There is also a need to validate existing questionnaires in terms of testing whether they actually measure what they purport to measure, with any acceptable degree of exactness. Therefore, it is suggested that, instead of doing extremely complicated multi-variate studies using statistical methods of forbidding complexity, researchers should spend at least some time trying to validate their basic data-gathering method. As is often said for factor analysis; 'Rubbish in, rubbish out'.

Stability of Accident Record

The review and analysis in Chapter 3 shows that we lack knowledge about stability of accident record over time regarding the following groups of drivers; motorcyclists, taxi and lorry drivers. There is also an almost total lack of research about stability between different road environments, vehicles and types of driving for the same individuals. Finally, how the amount and quality of exposure influences stability is virtually unknown.

It can also be noted that few researchers have studied what the effect of having an accident is on the probability of having another, or on the behaviour of the individual experiencing them, a research question that ties in with the stability problem. The stability meta-analysis would seem to indicate that we learn very little from experience, although the analyses undertaken in the original studies were correlational, and therefore do not really rule out that learning does occur. It is well known that experience of driving has an initial effect on risk of accident involvement, and it would be interesting to study whether some of this is due specifically to involvement in collisions. Today, the implicit view would seem to be that one learns by experience from driving, but not by having accidents. Whether this is actually so or not is not known.

In addition to the suggested research, it could also be studied whether there is fluctuation in accident risk over time for individuals, as suggested by Creswell and Froggatt (1963) in their 'spell' hypothesis. It can be noted that these 'spells' might be nothing else than fluctuations in accident proneness due to, for example, medical conditions, or other conditions that can change rapidly over time.

Determining Culpability

Very little research into drivers' responsibility for their collisions is available, when it comes to determining whether categorization is correct or not. As the correctness of the categorization is the basis for induced exposure, responsibility analysis, and to some degree, for studies that use culpable crashes only as outcome variable, this research is indeed very much needed.

First, some sort of theoretical development would be desirable. Presently, there is no basis for culpability judgements apart from the subjective view of the judge, and the responsibility analysis coding scheme. An organizing theory would facilitate the development of better-calibrated schemes, and possibly generate some testable hypotheses.

Meanwhile, the testing of the randomness assumption for non-culpable crashes in various populations is important. Given this type of research, it will at least be possible to calibrate coding schemes to achieve the right percentage of culpable involvements, although we would still only have a rather superficial understanding of the phenomenon as such.

It should be remembered that the problem of culpability is tied in with most of the other ones described in this book, and it will therefore tend to crop up whenever accident research is undertaken.

Accidents Versus Exposure

The research needed to explore the exposure problem is, first, studies using objectively gathered crash and mileage data. Second, if a non-proportional association is found, studies where drivers are allocated to groups experiencing different amounts of exposure, which is beyond their control, are required. If this also creates a non-proportionality (controlling for previous experience), it might thereafter be concluded that driving is indeed a skill that rapidly deteriorates and needs constant training. If this is the final conclusion, we will need to reconsider the previous thinking in traffic safety research about driving as an automated skill.

How to Construct an Accident Criterion

Although a good number of versions of the accident criterion have been used, no researcher seems to have studied them in their own right, that is, comparing different methods of calculation to see what type yields the best prediction by a certain independent variable. The knowledge in this area is close to zero, and the possibility of finding associations with accidents probably hinges very much upon how the criterion is constructed.

Here, specific research with the aim of showing that certain categories of accidents are better predicted by certain predictors cannot really be recommended, because every predictor will need its own specifically derived category of crashes as outcome variables, and there is therefore no one set of studies that can be made to solve the problem (possibly with the exception of culpability). Instead, it is recommended that all studies that try to predict individual differences in collision record use an 'All' crashes variable as well as a theoretically derived sub-set of these as outcome variables in parallel.

Alternative Outcome Variables

The basic need for alternative driving safety criteria is simple validation of the measure of interest. Such a recommendation can be found in any standard research methodological textbook. Validation can be done in the two ways; direct correlation with accidents, and secondary correlations for the effect sizes of the predictors between accidents and the alternative.

However, there does not really seem to be any use in actually recommending this type of research, because alternative variables will never be able to validly replace accidents as the measure of traffic safety. It is therefore recommended that researchers discontinue this detrimental practice and do something useful instead.

Discussion

Who Does Traffic Safety Research?

A few groups of traffic safety researchers seem to be discernible. First, there are the psychologists, sociologists and similar professionals, who are interested in the human mind in terms of fairly vague constructs, such as attitudes. Their main tool is the questionnaire, and it enables them to flood the journals with papers about traits and behaviours in any number of combinations.

Second, there are the technically and experimentally minded researchers, who use simulators, instrumented cars and similar set-ups. Here, the fairly strict designs and specific research questions can be contrasted with the often-questionable relevance for traffic safety (that is a lack of ecological validity). For example, although a study might find very nice differences in cognitive load between users and non-users of cell phones while driving, the extrapolation of this result to an actual effect upon accidents of cell phone use must be taken on faith.

Third, there are the medical researchers, who most often study risk as if it was an illness, with the methods common within their profession (most notably a preference for categorizing drivers into accident and no accident groups, and the use of logistic regression). In some respects, these authors make rigorous studies, applying the methodology they have learned within their main research field. Unfortunately, this is not always enough to avoid all the pitfalls described in this book.

The last observation leads to the general conclusion that there is a dire need to establish traffic safety as a research area in its own right, with a methodology that is suited to its needs. As it is, there seem to be no researchers who are especially trained for such work. Instead, a wide assortment of professions bring with them the tools of their original trade and try their hand at traffic research, which not only makes some of the undertakings very doubtful, but also makes the research overall very difficult to review. On the positive side, this pluralism of methodology counters conservatism and probably leads to new insights. What is lacking is a basis of tested methodology that can actually deliver reliable results. The themes discussed in this book could constitute such a basis, if properly researched.

What is 'Chance' in Accidents?

> Even in distributions based on chance it is possible to find that a few people are responsible for many accidents. McKenna, 1983, p. 67.

In the mid-age accident proneness debate, the focus was very much on the interpretation of accident distributions, mainly whether various empirically found distributions were different from the theoretically derived ones, and what this could mean. One basic notion was that of 'chance', as exemplified by this quote; 'It was not always recognized, however, that in the distribution of accidents among the members of a group, some individuals would have had more than others, *on the basis of chance alone.*' (McFarland and Moore, 1962, p. 377, italics added, see McFarland and Moore, 1957, for a similar statement).

What was the meaning of statements such as these? McKenna (1983) interpreted it as the similarity between the (probably very short-term) distribution of accidents and those found when processes that were not under any specific human control were studied, like the often-used analogy of shooting balls at pigeonholes.[2] This does seem to be a reasonable interpretation, and the similarity as such would not really seem to be a matter of controversy. However, if the similarity is taken as evidence that the accident distribution is 'random', very strange effects results, because the randomness assumption would seem to carry with it an implication of non-causality from the human side, as would seem to be implied by the quote from McFarland and Moore above.

This logic would seem to consist of a leap from what is mathematically considered to be random (not predictable?), to the actual world. But the similarity between theoretical and empirical distributions would seem to be a poor argument for claiming that the latter data are unpredictable. Actually, if a way was found of predicting the accident distribution, would it then cease to be random in the eyes of the statistician?

Also, being 'random' does not mean that the events that create the distribution are without cause. Actually, it means that the mathematician is not an empirical researcher, and has not investigated the reasons properly, and that the reasoning about randomness is directly misleading when applied to a real-world phenomenon.

---

2 This analogy is largely uninterpretable today, but probably refers to the British habit of keeping pigeons in nests in walls, which were simply a lot of holes into the wall. The analogy is about this; if you throw a ball at a wall with pigeonholes, it will end up in a certain area, and if you repeat this a large number of times, letting some sort of (not very accurate) machine do the throwing, the number of balls ending up in various parts of the wall will be 'chance distributed'.

It can also be noted that the discussion about the statistics of accident proneness never lead anywhere, in terms of predictions or empirical research. It would therefore seem like the statistical reasoning regarding chance and distributions should be considered irrelevant when phenomena like accidents are studied, if the aim is to gain any practically useful knowledge, for example, about why accidents happen and how this can be predicted.

The Effects (of Acceptance) of Bad Research

As noted in the introduction, research into individual differences in traffic safety has failed in its main task; the prediction of who is a dangerous driver. This should now be even more apparent, after the evidence for various shortcomings of the available research have been shown. However, it can also be pointed out that there are many who claim that they can actually predict such differences, and sell these services at a fairly high price (the Vienna Driving Test discussed in Chapter 8 is one example). Sometimes, their claims are even accepted by other researchers, as Evans (2004) apparently accepted the Virtual Fleet Manager (VFM) system in his influential book *Traffic Safety*. Unfortunately, the VFM does not really have any predictive power vis-à-vis accidents, as can even be shown using their own data (af Wåhlberg, 2007c). The end result is not only that government organizations and private companies pay for worthless tests; this practice is also a hindrance when it comes to researching and marketing more effective methods.

Yet another effect of bad research is that some groups may be erroneously treated because they have mistakenly been identified by some sort of method as more or less accident prone than others. In Sweden, the Road Administration suggested in 2007 that youngsters with an ADHD diagnosis should not be allowed to get a driving license, based upon 'the solid evidence' for an elevated risk for this group. As shown in Chapter 2, this evidence is mainly self-reported and probably to a fair degree due to social desirability effects.

Similarly, governments around the world keep using unsubstantiated methods for granting drivers their driving licences, thus allowing drivers that could probably have been identified as dangerous to roam free until they kill someone. Driving tests simply have no proven validity for the use to which they are put (the use of graduated licensing schemes are a different matter; they do not rely on the tests for their effect, but the experience gained under supervision).

In the end, research into individual differences could provide so much better knowledge and methods for selection, surveillance and training of drivers than what is currently used, thus improving the traffic safety situation greatly. Today, this type of research is hardly helpful for the practitioners, but rather the opposite.

The State of the Art

> The data for this study were obtained from a larger sample of 247 subjects who
> had been given LSD as part of psychotherapy or under experimental conditions...
> Jamison and McGlothlin, 1973, p. 123.

Have researchers grown tired of trying to predict accident involvement? Have we
turned to other venues because no great success ever ensued? It can be noted that
in the call for papers of the 4th International Conference on Traffic and Transport
Psychology (2008), none of the suggested themes included accidents, safety or
risk. Based upon the analyses undertaken in the present work, such a trend is very
worrying, because the presently available results on these topics can to a large
degree not be trusted.

Has traffic safety research become post-modernistic? The widespread use of
self-reported data and poorly validated variables would seem to indicate that there
is little interest in objectivity. Instead, much energy is put into 'theoretical models'[3]
and ad hoc questionnaires, with little predictive power, and even less validity and
practical applicability.

The results taken as truth within traffic safety research today can therefore to
a large degree be mistrusted. Here, some problems that have not been specifically
treated in the present book will be briefly analysed, and the basis for the general
thinking that seems to exist about them questioned. This is an undertaking that has
the aim of showing how a more critical view of the traffic safety research can reach
quite different conclusions about phenomena that are today seen as unproblematic
(in terms of being understood and having known remedies), both by researchers
and practitioners.

Some Thought Experiments

1. Alco-Locks

Once upon a time on a traffic conference I met a (Swedish) high-ranking police
officer who taught courses in traffic safety at an international level. He was very
positive to the use of alco-locks on cars, especially for known drunk drivers
(meaning that they would only be able to drive their cars when they were sober).
In his opinion, this would solve the whole problem of alcohol-related deaths on
the road.

---

3 The common use of the bastard term 'theoretical model' (for example, Åberg and
Rimmö, 1998; Gonzales, Dickinson, DiGuiseppi and Lowenstein, 2005) is in itself an
indication of a low degree of scientific quality. By using this strange term, the authors try
to have the credit of working with 'theory', while avoiding the demands of real scientific
theory (af Wåhlberg, 2001).

I, on the other hand, asked him whether he had considered the following scenario:

'We do not really know how alcoholics drive when they are under the influence. It is possible that at least some drive slower and more carefully,[4] avoiding busy roads and so on when they know they are drunk. This may counteract their otherwise lowered performance and actually make them safer. On the other hand, alcoholics tend to be come from groups with high accident rates in the first place, also when sober.

So, what happens when you force a number of drivers to drive, not when they are drunk and careful, but hung over and maybe not so careful? The actual consequence of alco-locks may be an increase in deaths on the road.' At this point, the high-ranking police officer called me a drug liberal and walked away.

It is my belief that this policeman believed, as many other, that alcohol is one of the main killers on the road, and that removing drunk, alcoholic drivers would decrease the road kill considerably. It is also my belief that he should have read Zylman's (1974) critical review of the beliefs and the evidence concerning alcohol and highway deaths, or Hedlund's (1994) paper, aptly titled 'If they didn't drink, would they crash anyway?'. His conclusion is well worth noting: 'Would they crash anyway? – some, but far fewer' (p. 124). The reason for this was apparently the similarity between drunk crash-involved drivers and other crashers. Drunks do not really need to be drunk to crash, because there is something about them which make them more liable both to drink and to crash, whether in combination or not, as noted by Smart and Schmidt (1969). It could therefore be questioned whether the number of road accidents that are usually considered to be due to drunk driving is really as high as some say. Maybe the drinking is just an auxiliary phenomenon that is noted because it is easily measured, while the real reasons are not? Today, it seems we do not really know this, despite the questions having been posed as early as decades ago.

2. Driver Feedback

For decades, all vehicle engines have become more powerful, and the vehicles themselves designed for forever higher speeds. It is also fairly apparent that people use the opportunities at their disposal in ways that are not compatible with safety. It might then be asked; why do people drive in such dangerous ways? One of the answers is probably that they do not experience any feeling of risk, due to a very high level of comfort, that is low level of noise, vibration etc. But what would happen if they were given feedback that indicated high speed? Frank McKenna described this type of general idea in a conversation in 2005, and said that he had been trying for years to tell vehicle makers the safety importance of such factors, to no avail.

---

4   This assumption is not borne out by research, at least not for a non-alcoholic population (Bates and Blakely, 1999).

A more specific method, which I suggested, would be to actually manipulate the driving seat in the same way as simulators do; by tilting it in various directions, an artificial feeling of acceleration could be created. Now, if drivers were not told about this (and the tilting could be done without being visible to passengers), it would seem very probable that those who like the thrill of acceleration could have their fix without the actual thing, a kind of placebo driving.

3. Increased Traffic Safety Over Time

What has caused the general decline in accident rates in the industrialized world during the last hundred years? Most practitioners and researchers would probably say that improvements of roads, training, medical aid and vehicles are the main reasons (some car manufacturers claim its all about the vehicles). However, there is one possible reason, which no one seems to think of; sheer experience. Today, every child in motorized countries is exposed to vehicles and their dangers, and humans do learn from experience (even though this may at times be doubted). It can therefore be assumed that children exposed to cars will learn how to avoid the dangers to a degree that is positively related to their amount of exposure. However, knowledge and (safety) practices are also handed down through the generations, and a long-standing vehicle prevalence (or dependency) will therefore have generated more of these commodities.

The conclusion from this experiment in thought is that we may have over-estimated the effects of our various traffic safety measures, when we look at them on a national level and over several decades. Unfortunately, this proposition is hard to test. What would be needed is a community where cars are introduced, but no regulations, or changes in roads, vehicle construction and training. If such a population could be found, it is predicted that their accident rate would diminish over time, despite the lack of 'improvement' in various safety technologies.

Final Words

> Two men saying they are Jesus. One of them must be wrong. Knopfler, 1982, t. 3.

This book has attitude. It can be summarized as disenchantment with the discrepancy between available results and actual practice concerning the methods used in traffic safety research, especially concerning individual differences, but also the lack of methodological discussion within this field. Unfortunately, the situation would seem to have grown worse with time. At least the early researchers could claim ignorance concerning a number of methodological pitfalls. This should not be the fact today, but on the contrary, rampant ignorance and false beliefs seem to be on the rise, while the quality of research plunges.

Many of the statements in this book have probably been offensive to a fair number of people. This was the intention, and was also deemed as necessary. Unfortunately, the things that have been stated here could never be said in journal papers, because this would offend someone, and editors seem to be very reluctant to publish anything that goes against the prevailing norm. But what is offensive to one person is sweet music to another. Personally, I get offended by the many casual claims made by various researchers about effects, validities and interpretations. To me, it is bad science, and thus very offensive.

The present book uses a lot of references. Actually, in all places where statements of facts have been made, literature has been referred to show where the information has been gathered. In some places, there have been attempts to review all available papers. Naturally, given the fairly long and prolific history of transport and accident research, these attempts have probably failed when it comes to being all encompassing. However, if any reader feels that some important (or even worse, their own!) work has been left out, I will gladly receive any communication about this, and try to include it in my future meanderings on this subject.

It is my hope that this book will create a discussion and a more critical research climate, and influence researchers to test what the consequences are of the many methodological pitfalls indicated here. I simply challenge the majority of researchers in individual differences in traffic safety to improve upon their methods. If not, a second edition of this book will certainly be published, which is something that I am not really hoping for. An improved state of the methodological art would be so much more preferable.

Returning to the introduction of the book, it should be remembered that research into individual differences in traffic accident record has not been a success. Although, this book has shown that the state of research into individual differences in traffic safety is much worse than usually assumed, this fact also brings with it a hope of improvement. Looking on the bright side of life, one thing we know for certain is that people differ a lot in their accident proneness. In principle, this should be possible to predict, and finally, research into individual differences should be able to significantly contribute to traffic safety.

References

Aarts, L., and van Schagen, I. (2006). Driving speed and the risk of road crashes: A review. *Accident Analysis and Prevention, 38,* 215–224.

Adams, J. R. (1970). Personality variables associated with traffic accidents. *Behavioral Research in Highway Safety, 1,* 3–18.

Adams-Guppy, J. R., and Guppy, A. (1995). Speeding in relation to perceptions of risk, utility and driving style by British company car drivers. *Ergonomics, 38,* 2525–2535.

Adams-Guppy, J. R., and Guppy, A. (2003). Truck driver fatigue risk assessment and management: A multinational survey. *Ergonomics, 46,* 763–779.

Adelstein, A. M. (1952). Accident proneness; a criticism of the concept based on analysis of shunter's accidents. *Journal of the Royal Statistical Society, 115,* 111–116.

Adler, A. (1941). The psychology of repeated accidents in industry. *American Journal of Psychiatry, 98,* 99–101.

Albery, I. P., Strang, J., Gossop, M., and Griffiths, P. (2000). Illicit drugs and driving: Prevalence, beliefs and accident involvement among a cohort of current out-of-treatment drug users. *Drug and Alcohol Dependence, 58,* 197–204.

Aldrich, M. S. (1989). Automobile accidents in patients with sleep disorders. *Sleep, 12,* 487–494.

Alexander, C. (1953). Psychological tests for drivers at the McLean Trucking Company. *Traffic Quarterly, 7,* 186–197.

Alm, H., and Nilsson, L. (1994). Changes in driver behaviour as a function of handsfree mobile phones – a simulator study. *Accident Analysis and Prevention, 26,* 441–451.

Alonso, A., Laguna, S., and Seguí-Gomez, M. (2006). A comparison of information on motor vehicle crashes as reported by written or telephone interviews. *Injury Prevention, 12,* 117–120.

Alsop, J., and Langley, J. (2001). Under-reporting of motor vehicle traffic crash victims in New Zealand. *Accident Analysis and Prevention, 33,* 353–359.

Alvarez, F. J., and Fierro, I. (2008). Older drivers, medical condition, medical impairment and crash risk. *Accident Analysis and Prevention, 40,* 55–60.

Anastasi, A. (1988). *Psychological Testing.* 6th ed. New York: MacMillan.

Anstey, K. J., Wood, J., Lord, S., and Walker, J. G. (2005). Cognitive, sensory and physical factors enabling driving safety in older adults. *Clinical Psychology Review, 25,* 45–65.

Aptel, I., Salmi, L. R., Masson, F., Bourde, A., Henrion, G., and Erny, P. (1999). Road accident statistics: Discrepancies between police and hospital data in a French island. *Accident Analysis and Prevention, 31,* 101–108.

Arbous, A. G., and Kerrich, J. E. (1951). Accident statistics and the concept of accident proneness. *Biometrics, 7,* 340–432.

Arendasy, M. E., Hergovich, A., Sommer, M., and Bognar, B. (2005). Dimensionality and construct validity of a video-based, objective personality test for the assessment of willingness to take risks in road traffic. *Psychological Reports, 97,* 309–320.

Armitage, C. J., and Conner, M. (2001). Efficacy of theory of planned behaviour: A meta-analytical review. *British Journal of Social Psychology, 40,* 471–499.

Armstrong, J. L., and Whitlock, F. A. (1980). Mental illness and road traffic accidents. *Australian and New Zealand Journal of Psychiatry, 14,* 53–60.

Arthur, W. Jr. (1991). *Individual Differences in the Prediction and Training of Complex Perceptual-Motor Skill Tasks: The Development and Validation of The Computer-Administered Test of Visual Selective Attention.* Technical report. Texas A&M University.

Arthur, W. Jr., Bell, S. T., Edwards, B. D., Day, E. A., Tubre, T. C., and Tubre, A. H. (2005). Convergence of self-report and archival motor vehicle crash involvement data: A two-year longitudinal follow up. *Human Factors, 47,* 303–313.

Arthur, W. Jr., and Doverspike, D. (1992). Locus of control and auditory selective attention as predictors of driving accident involvement: A comparative longitudinal investigation. *Journal of Safety Research, 23,* 73–80.

Arthur, W. Jr., and Doverspike, D. (2001). Predicting motor vehicle crash involvement from a personality measure and a driving knowledge test. *Journal of Prevention and Intervention in the Community, 22,* 35–42.

Arthur, W. Jr., and Graziano, W. G. (1996). The Five-factor model, conscientiousness, and driving accident involvement. *Journal of Personality, 64,* 593–618.

Arthur, W. Jr., Strong, M. H., and Williamson, J. (1994). Validation of a visual attention test as a predictor of driving accident involvement. *Journal of Occupational and Organizational Psychology, 67,* 173–182.

Arthur, W. Jr., Tubre, T. C., Day, E., Sheehan, M. K., Sanchez-Ku, M. L., Paul, D. S., Paulus, L. E., and Archuleta, K. D. (2001). Motor vehicle crash involvement and moving violations: Convergence of self-report and archival data. *Human Factors, 43,* 1–11.

Asbridge, M., Poulin, C., and Donato, A. (2005). Motor vehicle collision risk and driving under the influence of cannabis: Evidence from adolescents in Atlantic Canada. *Accident Analysis and Prevention, 37,* 1025–1034.

de Assis Viegas, C., and de Oliviera, H. W. (2006). Prevalence of risk factors for obstructive sleep apnea syndrome in interstate bus drivers. *Journal of Brazilian Pneumology, 32,* 144–149.

Austin, K. (1995a). The identification of mistakes in road accident records: Part I, Locational variables. *Accident Analysis and Prevention, 27,* 261–276.

Austin, K. (1995b). The identification of mistakes in road accident records: Part II, Casualty variables. *Accident Analysis and Prevention, 27,* 277–282.

Avolio, B. J., Kroeck, K. G., and Panek, P. E. (1985). Individual differences in information-processing ability as a predictor of motor vehicle accidents. *Human Factors, 27,* 577–589.

Babarik, P. (1968). Automobile accidents and driver reaction pattern. *Journal of Applied Psychology, 52,* 49–54.

Bach, H., Bickel, H., and Biehl, B. (1975). Validierung von Testverfahren zur Fahrer-Auslese am Unfallkriterium. *Zeitschrift fur Verkehrssicherheit, 21,* 27–38.

Baldock, M. R., Mathias, J. L., McLean, A. J., and Berndt, A. (2006). Self-regulation of driving and its relationship to driving ability among older adults. *Accident Analysis and Prevention, 38,* 1038–1045.

Ball, K., and Owsley, C. (1991). Identifying correlates of accident involvement for the older driver. *Human Factors, 33,* 583–595.

Ball, K., Owsley, C., Sloane, M. E., Roenker, D. L., and Bruni, J. (1993). Visual attention problems as predictor of vehicle crashes in older drivers. *Investigative Ophthalmology and Visual Science, 34,* 3110–3123.

Ball, K., Owsley, C., Stalvey, B., Roenker, D. L., Sloane, M. E., and Graves, M. (1998). Driving avoidance and functional impairment in older drivers. *Accident Analysis and Prevention, 30,* 313–322.

Banks, W., Shaffer, J., Masemore, W., Fisher, R., Schmidt, C., and Zlotowitz, H. (1977). The relationship between previous driving record and driver culpability in fatal, multiple-vehicle collisions. *Accident Analysis and Prevention, 9,* 9–14.

Barbone, F., McMahon, A. D., Davey, P. G., Morris, A. D., Reid, I. C., McDevitt, D. G., and McDonald, T. M. (1998). Association of road-traffic accidents with benzodiazepine use. *Lancet, 352,* 1331–1336.

Barger, L. K., Cade, B. E., Ayas, N. T., Cronin, J. W., Rosner, B., Speizer, F. E., and Czeisler, C. A. (2005). Extended work shifts and the risk of motor vehicle crashes among interns. *New England Journal of Medicine, 352,* 125–134.

Barkley, R. A. (2006). Driving risks in adults with ADHD: Yet more evidence and a personal story. *The ADHD Report, 14,* 1–10.

Barkley, R. A., and Cox, D. (2007). A review of driving risks and impairments associated with attention-deficit/hyperactivity disorder and the effects of stimulant medication on driving performance. *Journal of Safety Research, 38,* 113–128.

Barkley, R. A., Fischer, M., Edelbrock, C. S., and Smallish, L. (1990). The adolescent outcome of hyperactive children diagnosed by research criteria: I. An 8-year propspective follow-up study. *Journal of the American Academy of Child and Adolescent Psychiatry, 29,* 546–557.

Barkley, R. A., Guevremont, D. C., Anastopoulos, A. D., DuPaul, G. J., and Shelton, T. L. (1993). Driving-related risks and outcomes of attention deficit hyperactivity disorder in adolescents and young adults: A 3- to 5-year follow-up survey. *Pediatrics, 92,* 212–218.

Barkley, R. A., Murphy, K. R., DuPaul, G. J., and Bush, T. (2002). Driving in young adults with attention deficit hyperactivity disorder: Knowledge, performance, adverse outcomes, and the role of executive functioning. *Journal of the International Neuropsychological Society, 8,* 655–672.

Barkley, R. A., Murphy, K. R., and Kwasnik, D. (1996). Motor vehicle driving competencies and risks in teens and young adults with attention deficit hyperactivity disorder. *Pediatrics, 98,* 1089–1095.

Barkley, R. A., Murphy, K. R., O'Connell, T., Anderson, D., and Connor, D. F. (2006). Effects of two doses of alcohol on simulator driving performance in adults with attention-deficit/hyperactivity disorder. *Neuropsychology, 20,* 77–87.

Barrett, G. V., and Thornton, C. L. (1968). Relationship between perceptual style and driver reaction to an emergency situation. *Journal of Applied Psychology, 52,* 169–176.

Barrett, G. V., Thornton, C. L., and Cabe, P. A. (1969). Relation between embedded figures test performance and simulator behaviour. *Journal of Applied Psychology, 53,* 253–254.

Barrick, M. R., and Mount, M. K. (1996). Effects of impression management and self-deception on the predicitive validity of personality constructs. *Journal of Applied Psychology, 81,* 261–272.

Bates, M. N., and Blakely, T. A. (1999). Role of cannabis in motor vehicle crashes. *Epidemiologic Reviews, 21,* 222–232.

Baughan, C. J., and Sexton, B. (2001). Do driving test errors predict accidents? Yes and No. *Behavioural Research in Road Safety XI,* 252–268. London: Department for Transport.

Beamish, J. J., and Malfetti, J. L. (1962). A psychological comparison of violator and non-violator automobile drivers in the 16 to 19 year age group. *Traffic Safety Research Review, 6,* 12–15.

Beck, K. H., Shattuck, T., and Raleigh, R. (2001). Parental predictors of teen driving risk. *American Journal of Health Behavior, 25,* 10–20.

Beck, K. H., Wang, M. Q., and Mitchell, M. M. (2006). Concerns, dispositions and behaviors of aggressive drivers: What do self-identified aggressive drivers believe about traffic safety? *Journal of Safety Research, 37,* 159–165.

Beck, K. H., Yan, F., and Wang, M. Q. (2007). Cell phone users, reported crash risk, unsafe driving behaviors and dispositions: A survey of motorists in Maryland. *Journal of Safety Research, 38,* 683–688.

Bedard, M., Dubois, S., and Weaver, B. (2007). The impact of cannabis on driving. *Canadian Journal of Public Health, 98,* 6–11.

Begg, D. J., Langley, J. D., and Williams, S. M. (1999). Validity of self-reported crashes and injuries in a longitudinal study of young adults. *Injury Prevention, 5,* 142–144.

Beirness, D. J., and Simpson, H. M. (1988). Lifestyle correlates of risky driving and accident involvement among youth. *Alcohol, Drugs and Driving, 4,* 193–204.

Bener, A., Murdoch, J. C., Achan, N. V., Karama, A. H., and Sztriha, L. (1996). The effect of epilepsy on road traffic accidents and casualties. *Seizure, 5,* 215–219.

Bener, A., Özkan, T., and Lajunen, T. (2008). The Driver Behaviour Questionnaire in Arab Gulf countries: Qatar and United Arab emirates. *Accident Analysis and Prevention, 40,* 1411–1417.

Bernacki, E. J. (1976). Accident proneness or accident liability: Which model for industry? *Connecticut Medicine, 40,* 535–538.

Bingham, C. R., and Shope, J. T. (2005). Adolescent predictors of traffic crash patterns from licensure into early young adulthood. *Annual Proceedings Association for the Advancement of Automotive Medicine, 49,* 237–252.

Blanchard, E. B., Barton, K. A., and Malta, L. (2000). Psychometric properties of a measure of aggressive driving: The Larson Driver's Stress Profile. *Psychological Reports, 87,* 881–892.

Blasco, R. D., Prieto, J. M., and Cornejo, J. M. (2003). Accident probability after accident occurence. *Safety Science, 41,* 481–501.

Blockey, P. N., and Hartley, L. R. (1995). Aberrant driving behaviour: Errors and violations. *Ergonomics, 38,* 1759–1771.

Blows, S., Ameratunga, S., Ivers, R. Q., Lo, S. K., and Norton, R. (2005). Risky driving habits and motor vehicle driver injury. *Accident Analysis and Prevention, 37,* 619–624.

Blum, M. L., and Mintz, A. (1951). Correlation versus curve fitting in research on accident proneness: Reply to Maritz. *Psychological Bulletin, 48,* 413–418.

Bouillon, L., Mazer, B., and Gelinas, I. (2006). Validity of the cognitive behavioral driver's inventory in predicting driving outcome. *American Journal of Occupational Therapy, 60,* 420–427.

Boyce, T. E., and Geller, E. S. (2002). An instrumented vehicle assessment of problem behavior and driving style: Do younger males really take more risks? *Accident Analysis and Prevention, 34,* 51–64.

Boyle, A. J. (1980). 'Found experiments' in accident research: Report of a study of accident rates and implications for future research. *Journal of Occupational Psychology, 53,* 53–64.

Brandaleone, H., and Flamm, E. (1955). Psychological testing: Effect on the accident frequency of bus operators. *Industrial Medicine and Surgery, 24,* 296–298.

Brookhuis, K. A., and de Waard, D. (1993). The use of psychophysiology to assess driver status. *Ergonomics, 36,* 1099–1110.

Brorsson, B., Rydgren, H., and Ifver, J. (1993). Single-vehicle accidents in Sweden: A comparative study of risk and risk factors by age. *Journal of Safety Research, 24,* 55–65.

Brown, C. W., and Ghiselli, E. E. (1948a). Accident proneness among streetcar motormen and motor coach operators. *Journal of Applied Psychology, 32,* 20–23.

Brown, C. W., and Ghiselli, E. E. (1948b). Factors related to the proficiency of motor coach operators. *Journal of Applied Psychology, 31,* 477–479.

Brown, P. L., and Berdie, R. F. (1960). Driver behavior and scores on the MMPI. *Journal of Applied Psychology, 44,* 18–21.

Brown, S. L., and Cotton, A. (2003). Risk-mitigating beliefs, risk estimates, and self-reported speeding in a sample of Australian drivers. *Journal of Safety Research, 34,* 183–188.

Brown, T. D. (1976). Personality traits and their relationship to traffic violations. *Perceptual and Motor Skills, 42,* 467–470.

Bull, J. P., and Roberts, B. J. (1973). Road accident statistics – a comparison of police and hospital information. *Accident Analysis and Prevention, 5,* 45–53.

Burg, A. (1967). *The Relationship Between Vision Test Scores and Driving Record: General Findings.* Report No. 67–24. California: Department of Motor Vehicles.

Burg, A. (1970). The stability of driving record over time. *Accident Analysis and Prevention, 2,* 57–65.

Burg, A. (1974). *Traffic Violations in Relation to Driver Characteristics and Accident Frequency.* Los Angeles: Institute of Transportation and Traffic Engineering, UCLA.

Burger, J. M. (1981). Motivational bias in the attribution of responsibility for an accident: A meta-analysis of the Defensive-Attribution hypothesis. *Psychological Bulletin, 90,* 496–512.

Burns, P. C., and Wilde, G. J. (1995). Risk taking in male taxi drivers: Relationships among personality, observational data and driver records. *Personality and Individual Differences, 18,* 267–278.

Bygren, L. D. (1974). The driver's exposure to risk of accident. *Scandinavian Journal of Social Medicine, 2,* 49–65.

Caird, J. K., and Kline, T. J. (2004). The relationships between organizational and individual variables to on-the-job driver accidents and accident-free kilometres. *Ergonomics, 47,* 1598–1613.

Cameron, C. (1975). Accident proneness. *Accident Analysis and Prevention, 7,* 49–53.

Campbell, D. T., and Fiske, D. W. (1959). Convergent and discriminant validation by the multitrait-multimethod matrix. *Psychological Bulletin, 56,* 81–105.

Campbell, R. E., and King, L. E. (1970). The traffic conflicts technique applied to rural intersections. *Accident Analysis and Prevention, 2,* 209–221.

Carr, B. R. (1969). A statistical analysis of rural Ontario traffic accidents using induced exposure data. *Accident Analysis and Prevention, 5,* 343–357.

Carty, M., Stough, C., and Gillespie, N. (1998). The psychological predictors of work accidents and driving convictions in the transport industry. *Safety Science Monitor, 3,* 1–13.

Cash, W. S., and Moss, A. J. (1972). *Optimum Recall Period for Reporting Persons Injured in Motor Vehicle Accidents.* Vital and Health Statistics Series 2 No 50. DHEW Publication No 72-1050. US Department of Health, Education, and Welfare.

Cassel, W., Ploch, T., Becker, C., Dugnus, D., Peter, J. H., and von Wichert, P. (1996). Risk of traffic accidents in patients with sleep-disordered breathing: Reduction with nasal CPAP. *European Respiratory Journal, 9,* 2606–2611.

Cation, W. L., Mount, G. E., and Brenner, R. (1951). Variability of reaction time and susceptibility to automobile accidents. *Journal of Applied Psychology, 35,* 101–107.

Cellar, D. F., Nelson, Z. C., and York, C. M. (2000). The five-factor model and driving behavior: Personality and involvement in vehicular accidents. *Psychological Reports, 86,* 454–456.

Cercarelli, L. R., Arnold, P. K., Rosman, D. L., Sleet, D., and Thornett, M. L. (1992). Travel exposure and choice of comparison crashes for examining motorcycle conspicuity by analysis of crash data. *Accident Analysis and Prevention, 23,* 363–368.

Cerrelli, E. C. (1973). Driver exposure the indirect approach for obtaining relative measures. *Accident Analysis and Prevention, 5,* 147–156.

Chaloupka, C., and Risser, R. (1995). Don't wait for accidents – possibilities to assess risk in traffic by applying the 'Wiener Fahrprobe'. *Safety Science, 19,* 137–147.

Chandraratna, S., Stamatiadis, N., and Stromberg, A. (2006). Crash involvement of drivers with multiple crashes. *Accident Analysis and Prevention, 38,* 532–541.

Chapman, P., and Underwood, G. (2000). Forgetting near-accidents: The roles of severity, culpability and experience in the poor recall of dangerous driving situations. *Applied Cognitive Psychology, 14,* 31–44.

Chapman, R. (1973). The concept of exposure. *Accident Analysis and Prevention, 5,* 95–110.

Chen, W., Cooper, P., and Pinili, M. (1995). Driver accident risk in relation to the penalty point system in British Columbia. *Journal of Safety Research, 26,* 9–18.

Chin, H.-C., and Quek, S.-T. (1997). Measurement of traffic conflicts. *Safety Science, 26,* 169–185.

Chipman, M. L. (1982). The role of exposure, experience and demerit point levels in the risk of collision. *Accident Analysis and Prevention, 14,* 475–483.

Chipman, M. L., MacDonald, S., and Mann, R. E. (2003). Being 'at fault' in traffic crashes: Does alcohol, cannabis, cocaine, or polydrug abuse make a difference? *Injury Prevention, 9,* 343–348.

Chipman, M. L., MacGregor, C. G., Smiley, A. M., and Lee-Gosselin, M. (1992). Time vs. distance as measures of exposure in driving surveys. *Accident Analysis and Prevention, 24,* 679–684.

Chipman, M. L., MacGregor, C. G., Smiley, A. M., and Lee-Gosselin, M. (1993). The role of exposure in comparisons of crash risk among different drivers and driving environments. *Accident Analysis and Prevention, 25,* 207–211.

Chu, B. Y., and Nunn, G. E. (1976). An analysis of the decline in California traffic fatalities during the energy crisis. *Accident Analysis and Prevention, 8,* 145–150.

Clarke, D. D., Ward, P. J., Bartle, C., and Truman, W. (2006). Young driver accidents in the UK: The influence of age, experience, and time of day. *Accident Analysis and Prevention, 38,* 871–878.

Clarke, D. D., Ward, P. J., and Jones, J. (1998). Overtaking road-accidents: Differences in manoeuvre as a function of driver age. *Accident Analysis and Prevention, 30,* 455–467.

Clarke, S. (2006). Contrasting perceptual, attitudinal and dispositional approaches to accident involvement in the workplace. *Safety Science, 44,* 537–550.

Clarke, S., and Robertson, I. T. (2005). A meta-analytic review of the Big Five personality factors and accident involvement in occupational and non-occupational settings. *Journal of Occupational and Organizational Psychology, 78,* 355–376.

Clement, R., and Jonah, B. A. (1984). Field dependence, sensation seeking and driving behaviour. *Personality and Individual Differences, 5,* 87–93.

Cliaoutakis, J. E., Demakakos, P., Tzamalouka, G., Bakou, V., Koumaki, M., and Darviri, C. (2002). Aggressive behavior while driving as predictor of self-reported car crashes. *Journal of Safety Research, 33,* 431–443.

Cobb, P. (1940). The limit of usefulness of accident rate as a measure of accident proneness. *Journal of Applied Psychology, 24,* 154–159.

Cobliner, W. G., and Shatin, L. (1962). An adaptional perspective on the traffic accident. *Traffic Safety Research Review, 6,* 13–15.

Conner, M., and Lai, F. (2005). *Evaluation of the Effectiveness of the National Driver Improvement Scheme.* Road Safety Report No. 64. London: Department for Transport.

Connor, J., Whitlock, G., Norton, R., and Jackson, R. (2001). The role of sleepiness in car crashes: A systematic review of epidemiological studies. *Accident Analysis and Prevention, 33,* 31–41.

Consiglio, W., Driscoll, P., Witte, M., and Berg, W. P. (2003). Effect of cellular telephone conversations and other potential interference on reaction time in a braking response. *Accident Analysis and Prevention, 35,* 495–500.

Cooper, P. J. (1997). The relationship between speeding behavior (as measured by violation convictions) and crash involvement. *Journal of Safety Research, 28,* 83–95.

Cooper, P. J., Tallman, K., Tuokko, H., and Beattie, B. (1993). Vehicle crash involvement and cognitive deficit in older drivers. *Journal of Safety Research, 24,* 9–17.

Corbett, C. (2001). Explanations for 'understating' in self-reported speeding behaviour. *Transportation Research Part F, 4,* 133–150.

Corbett, C., and Simon, F. (1992). *Unlawful Driving Behaviour: A Criminological Perspective.* TRL Contractor Report 301. Crowthorne: Transport Research Laboratory.

Cox, D. J., Penberthy, J. K., Zrebiec, J., Weinger, K., Aikens, J. E., Frier, B., Stetson, B., DeGroot, M., Trief, P., Schaechinger, H., Hermanns, N., Gonder-Frederick, L., and Clarke, W. (2003). Diabetes and driving mishaps: Frequency and correlations from a multinational survey. *Diabetes Care, 26,* 2329–2334.

Cox, D. J., Taylor, P., and Kovatchev, B. (1999). Driving simulation performance predicts future accidents among older drivers. *Journal of the American Geriatric Society, 47,* 381–382.

Crancer, A., and McMurray, L. (1968). Accident and violation rates of Washington's restricted drivers. *Journal of the American Medical Association, 205,* 272–276.

Crancer, A., and O'Neall, P. (1970). A record analysis of Washington drivers with license restrictions for heart disease. *Northwest Medicine, 69,* 409–416.

Crancer, A., and Quiring, D. L. (1969). The mentally ill as motor vehicle operators. *American Journal of Psychiatry, 126,* 807–813.

Crawford, P. L. (1960). Hazard exposure differentiation necessary for the identification of the accident-prone employee. *Journal of Applied Psychology, 44,* 192–194.

Crawford, W. A. (1971). Accident proneness: An unaffordable philosophy. *Medical Journal of Australia, 2,* 905–909.

Cresswell, W. L., and Froggatt, P. (1963). *The Causation of Bus Driver Accidents: An Epidemiological Study.* London: University Press.

Cummings, S. R., Nevitt, M. C., and Kidd, S. (1988). Forgetting falls. The limited accuracy of recall of falls in the elderly. *American Geriatric Society, 36,* 613–616.

Dahlen, E. R., Martin, R. C., Ragan, K., and Kuhlman, M. M. (2005). Driving anger, sensation seeking, impulsiveness, and boredom proneness in the prediction of unsafe driving. *Accident Analysis and Prevention, 37,* 341–348.

Dahlen, E. R., and Ragan, K. (2004). Validation of the Propensity for Angry Driving Scale. *Journal of Safety Research, 35,* 557–563.

Dahlen, E. R., and White, R. P. (2006). The Big Five factors, sensation seeking, and driving anger in the prediction of unsafe driving. *Personality and Individual Differences, 41,* 903–915.

Daigneault, G., Joly, P., and Frigon, J.-Y. (2002). Previous convictions or accidents and the risk of subsequent accidents of older drivers. *Accident Analysis and Prevention, 34,* 257–261.

Dalziel, J. R., and Job, R. F. (1997). Motor vehicle accidents, fatigue and optimism bias in taxi drivers. *Accident Analysis and Prevention, 29,* 489–494.

Darby, P., Murray, W., and Raeside, R. (2009). Applying online fleet driver assessment to help identify, target and reduce occupational road safety risks. *Safety Science, 47,* 436–442.

Davey, J. B. (1956). The effect of visual acuity on driving ability. *British Journal of Physiological Optics, 13,* 62–78.

Davis, D. R., and Coiley, P. A. (1959). Accident-proneness in motor-vehicle drivers. *Ergonomics, 2,* 239–246.

Davis, G. A., and Gao, Y. (1995). Statistical methods to support induced exposure analyses of traffic accident data. *Transportation Research Record, 1401,* 43–49.

Davison, P. A. (1978). The role of driver's vision in road safety. *Lighting Research and Technology, 10,* 125–139.

Davison, P. A. (1985). Inter-relationships between British drivers' visual abilities, age and road accident histories. *Ophthalmic and Physiological Optics, 5,* 195–204.

Decina, L. E., and Staplin, L. (1993). Retrospective evaluation of alternative criteria for older and younger drivers. *Accident Analysis and Prevention, 25,* 267–275.

Dee, T. S. (1998). Reconsidering the effects of seat belt laws and their enforcement status. *Accident Analysis and Prevention, 30,* 1–10.

Deery, H. A., and Fildes, B. N. (1999). Young novice driver subtypes: Relationship to high-risk behavior, traffic accident record, and simulator driving performance. *Human Factors, 41,* 628–643.

Deffenbacher, J. L., Filetti, L. B., Richards, T. L., Lynch, R. S., and Oetting, E. R. (2003). Characteristics of two groups of angry drivers. *Journal of Counseling Psychology, 50,* 123–132.

Deffenbacher, J. L., Huff, M. E., Lynch, R. S., Oetting, E. R., and Salvatore, N. F. (2000). Characteristics and treatment of high-anger drivers. *Journal of Counseling Psychology, 47,* 5–17.

Deffenbacher, J. L., Lynch, R. S., Deffenbacher, D. M., and Oetting, E. R. (2001). Further evidence of reliability and validity for the Driving Anger Expression Inventory. *Psychological Reports, 89,* 535–540.

Deffenbacher, J. L., Lynch, R. S., Filetti, L. B., Dahlen, E. R., and Oetting, E. R. (2003). Anger, aggression, risky behavior, and crash-related outcomes in three groups of drivers. *Behavior Research and Therapy, 41,* 333–349.

Deffenbacher, J. L., Lynch, R. S., Oetting, E. R., and Swaim, R. C. (2002). The Driving Anger Expression Inventory: A measure of how people express their anger on the road. *Behaviour Research and Therapy, 40,* 717–737.

Deffenbacher, J. L., Lynch, R. S., Oetting, E. R., and Yingling, D. A. (2001). Driving anger: Correlates and a test of of state-trait theory. *Personality and Indivdual Differences, 31,* 1321–1331.

Deffenbacher, J. L., Oetting, E. R., and Lynch, R. S. (1994). Development of a driving anger scale. *Psychological Reports, 74,* 83–91.

Deffenbacher, J. L., Oetting, E. R., Lynch, R. S., and Morris, C. D. (1996). The expression of anger and its consequences. *Behaviour Research and Therapy, 34,* 575–590.

Deffenbacher, J. L., Petrilli, R. T., Lynch, R. S., Oetting, E. R., and Swaim, R. C. (2003). The Driver's Angry Thoughts Questionnaire: A measure of angry cognitions when driving. *Cognitive Research and Therapy, 27,* 383–402.

Deffenbacher, J. L., Richards, T. L., Filetti, L. B., and Lynch, R. S. (2005). Angry drivers: A test of the state-trait theory. *Violence and Victims, 20,* 455–469.

Deffenbacher, J. L., White, G. S., and Lynch, R. S. (2004). Evaluation of two new scales assessing driver anger: The Driver Anger Expression Inventory and the Driver's Angry Thoughts Questionnaire. *Journal of Psychopathology and Behavioral Assessment, 26,* 87–99.

DePasquale, J. P., Geller, E. S., Clarke, S. W., and Littleton, L. C. (2001). Measuring road rage – development of the Propensity for Angry Driving Scale. *Journal of Safety Research, 32,* 1–16.

De Raedt, R., and Ponjaert-Kristoffersen, I. (2001). Predicting at-fault accidents of older drivers. *Accident Analysis and Prevention, 33,* 809–819.

Diges, M. (1988). Stereotypes and memory of real traffic accidents. In M. Gruneberg, P. Morris and R. Sykes (eds) *Practical Aspects of Memory: Memory in Everyday Life.* Chichester: John Wiley and Sons.

Dionne, G., Desjardins, D., Laberge-Nadeau, C., and Maag, U. (1995). Medical conditions, risk exposure, and truck drivers' accidents: An analysis with count data regression models. *Accident Analysis and Prevention, 27,* 295–305.

Dobson, A., Brown, W., Ball, J., Powers, J., and McFadden, M. (1999). Women drivers' behaviour, socio-demographic characteristics and accidents. *Accident Analysis and Prevention, 31,* 525–535.

Donovan, D. M., and Marlatt, G. A. (1982). Personality subtypes among driving-while-intoxicated offenders: Relationship to drinking behavior and driving risk. *Journal of Consulting and Clinical Psychology, 50,* 241–249.

Donovan, D. M., Queisser, H. R., Salzberg, P. M., and Umlauf, R. L. (1985). Intoxicated and bad drivers: Subgroups within the same population of high-risk drivers. *Journal of Studies on Alcohol, 46,* 375–382.

Donovan, D. M., Queisser, H. R., Umlauf, R. L., and Salzberg, P. M. (1986). Personality subtypes among driving-while-intoxicated offenders: Follow-up of subsequent driving records. *Journal of Consulting and Clinical Psychology, 54,* 563–565.

Dorn, L. (2005). Professional driver training and driver stress: Effects on simulated driving performance. In G. Underwood (ed.), *Traffic and Transport Psychology.* Elsevier.

Dorn, L., and af Wåhlberg, A. E. (2008). Work related road safety: An analysis based on UK bus driver performance. *Risk Analysis: An International Journal, 28,* 25–35.

Drake, C. A. (1940). Accident proneness: A hypothesis. *Character and Personality, 8,* 335–341.

Drummer, O. H., Gerostamoulos, J., Batziris, H., Chu, M., Caplehorn, J. R., Robertson, M. D., and Swann, P. (2004). The involvement of drugs in drivers of motor vehicles killed in Australian road traffic crashes. *Accident Analysis and Prevention, 36,* 239–248.

Dula, C. S., and Ballard, M. E. (2003). Development and evaluation of a measure of dangerous, aggressive, negative emotional, and risky driving. *Journal of Applied Social Psychology, 33,* 263–282.

Eadington, D. W., and Frier, B. M. (1989). Type 1 diabetes and driving experience: An eight year cohort study. *Diabetic Medicine, 6,* 137–141.

Edwards, D. S., and Hahn, C. P. (1980). A chance to happen. *Journal of Safety Research, 12,* 59–67.

Edwards, D. S., Hahn,. C. P., and Fleishman, E. D. (1977). Evaluation of laboratory methods for the study of driver behavior: Relations between simulator and street performance. *Journal of Applied Psychology, 62,* 559–566.

Elander, J., West, R. J., and French, D. J. (1993). Behavioral correlates of individual differences in road-traffic crash risk: An examination of methods and findings. *Psychological Bulletin, 113,* 279–294.

Elliott, D. (1987). Self-reported driving while under the influence of alcohol/drugs and the risk of alcohol/drug-related accidents. *Alcohol, Drugs and Driving, 3,* 31–43.

Elliott, M. A., Armitage, C. J., and Baughan, C. J. (2007). Using the theory of planned behaviour to predict observed driving behaviour. *British Journal of Social Psychology, 46,* 90–96.

Elliott, M. A., and Baughan, C. J. (2004). Developing a self-report method for investigating adolescent road user behaviour. *Transportation Research Part F, 7,* 373–393.

Engleman, H. M., Hirst, W. S., and Douglas, N. J. (1997). Under reporting of sleepiness and driving impairment in patients with sleep apnoea/hypopnoea syndrome. *Journal of Sleep Research, 6,* 272–275.

Evans, L. (2004). *Traffic Safety.* Bloomfield: Science Serving Society.

Evans, L., and Gerrish, P. H. (1996). Antilock brakes and risk of front and rear impact in two-vehicle crashes. *Accident Analysis and Prevention, 28,* 315–323.

Evans, R., Hansen, W., and Mittlemark, M. (1977). Increasing the validity of self-reports of behaviour in a smoking in children investigation. *Journal of Applied Psychology, 62,* 521–523.

Farmer, C. M., Braver, E. R., and Mitter, E. L. (1997). Two vehicle side impact crashes: The relationship of vehicle and crash characteristics to injury severity. *Accident Analysis and Prevention, 29,* 399–406.

Farmer, C. M., Retting, R. A., and Lund, A. K. (1999). Changes in motor vehicle occupant fatalities after repeal of the national maximum speed limit. *Accident Analysis and Prevention, 31,* 537–543.

Farmer, E., and Chambers, E. G. (1926). *A Psychological Study of Individual Differences in Accident Rates.* Industrial Health Research Board Report No 38. London: His Majesty's Stationery Office.

Farmer, E., and Chambers, E. G. (1939). *A Study of Accident Proneness Among Motor Drivers.* Industrial Health Research Board Report No 84. London: His Majesty's Stationery Office.

Farrand, P., and McKenna, F. P. (2001). Risk perception in novice drivers: The relationship between questionnaire measures and response latency. *Transportation Research Part F, 4,* 201–212.

Feldman, J. M., and Lynch, J. G. (1988). Self-generated validity and other effects of measurement on belief, attitude, intention, and behavior. *Journal of Applied Psychology, 73,* 421–435.

Feldman, P. J., Cohen, S., Doyle, W. J., Skoner, D. P., and Gwaltney, J. M. (1999). The impact of personality on the reporting of unfounded symptom and illness. *Journal of Personality and Social Psychology, 77,* 370–378.

Ferdun, G. S., Peck, R. C., and Coppin, R. S. (1967). The teen-aged driver: An evaluation of age, experience, driving exposure and driver training as they relate to driving record. *Highway Research Record, 163,* 31–53.

Fergenson, P. E. (1971). The relationship between information processing and driving accident and violation record. *Human Factors, 13,* 173–176.

Fergusson, D. M., and Horwood, L. J. (2001). Cannabis use and traffic accidents in a birth cohort of young adults. *Accident Analysis and Prevention, 33,* 703–711.

Fergusson, D. M., Horwood, L. J., and Boden, J. M. (2008). Is driving under the influence of cannabis becoming a greater risk to driver safety than drink driving? Findings from a longitudinal study. *Accident Analysis and Prevention, 40,* 1345–1350.

Fernandes, R., Job, R. F., and Hatfield, J. (2007). A challenge to the assumed generalizability of prediction and countermeasure for risky driving: Different factors predict different risky driving behaviors. *Journal of Safety Research, 38,* 59–70.

Ferrando, P. J. (2008). The impact of social desirability bias on the EPQ-R item scores: An item response theory analysis. *Personality and Individual Differences, 44,* 1748–1794.

Ferrie, J. E., Kivimäki, M., Head, J., Shipley, M. J., Vahtera, J., and Marmot, M. G. (2005). A comparison of self-reported sickness absence with absences recorded in employers' registers: Evidence from the Whitehall II study. *Occupational Environmental Medicine, 62,* 74–79.

Fhanér, G., and Hane, M. (1973). Seat belts: The importance of situational factors. *Accident Analysis and Prevention, 5,* 267–285.

Fife, D., and Cadigan, R. (1989). Regional variation in motor vehicle accident reporting: Findings from Massachusetts. *Accident Analysis and Prevention, 21,* 193–196.

Findley, L. J., Fabrizio, M., Knight, H., Norcross, B., LaForte, A., and Suratt, P. M. (1989). Driving simulator performance in patients with sleep apnea. *American Review of Respiratory Disease, 140,* 529–530.

Findley, L. J., Levinson, M. P., and Bonnie, R. J. (1992). Driving performance and automobile accidents in patients with sleep apnoea. *Clinics in Chest Medicine, 13,* 427–435.

Findley, L. J., Smith, C., Hooper, J., Dineen, M., and Suratt, P. M. (2000). Treatment with nasal CPAP decreases accidents in patients with sleep apnea. *American Journal of Respiratory and Critical Care Medicine, 161,* 857–859.

Findley, L. J., Unverzagt, M. E., and Suratt, P. M. (1988). Automobile accidents involving patients with obstructive sleep apnoea. *American Review of Respiration Disorder, 138,* 337–340.

Fischer, M., Barkley, R. A., Smallish, L., and Fletcher, K. (2007). Hyperactive children as young adults: Driving abilities, safe driving behavior, and adverse driving outcomes. *Accident Analysis and Prevention, 39,* 94–105.

Fitten, L., Perryman, K., Wilkinson, C., Little, R., Burns, M., Pachana, N., Mervis, J., Malmgren, R., Siembeda, D., and Ganzell, S. (1995). Alzheimer and vascular dementias. A prospective road and laboratory study. *Journal of the American Medical Association, 273,* 1360–1365.

Foley, D., Wallace, R., and Eberhard, J. (1995). Risk factors for motor vehicle crashes among older drivers in a rural community. *Journal of the American Geriatric Society, 43,* 776–781.

Forbes, T. W. (1939). The normal automobile driver as a traffic problem. *Journal of General Psychology, 20,* 471–474.

Forbes, T. W. (1957). Analysis of 'near-accident' reports. *Highway Research Board Bulletin, 152,* 23–37.

Foreman, E. I., Ellis, H. D., and Beavan, D. (1983). Mea culpa? A study of the relationships among personality traits, life-events and ascribed accident causation. *British Journal of Clinical Psychology, 22,* 223–224.

French, D. J., West, R. J., Elander, J., and Wilding, J. M. (1993). Decision-making style, driving style, and self-reported involvement in road traffic accidents. *Ergonomics, 36,* 627–644.

Fried, R., Petty, C. R., Surman, C. B., Reimer, B., Aleardi, M., Martin, J. M., Coughlin, J. F., and Biederman, J. (2006). Characterizing impaired driving in adults with attention-deficit/hyperactivity disorder: A controlled study. *Journal of Clinical Psychiatry, 6,* 567–574.

Friedland, R., Koss, E., Kumar, A., Gaine, S., Metzler, D., Haxby, J., and Moore, A. (1988). Motor vehicle crashes in dementia of the Alzheimer type. *Annals of Neurology, 24,* 782–786.

Froggatt P., and Smiley J. A. (1964). The concept of accident proneness: A review. *British Journal of Industrial Medicine, 21,* 1–12.

Fuller, R. G. (2005). Towards a general theory of driver behaviour. *Accident Analysis and Prevention, 37,* 461–472.

Furnham, A., and Saipe, J. (1993). Personality correlates of convicted drivers. *Personality and Individual Differences, 14,* 329–336.

Galovski, T. E., and Blanchard, E. B. (2002). The effectiviness of a brief psychological intervention on court-referred and self-referred aggressive drivers. *Behaviour Research and Therapy, 40,* 1385–1402.

Galski, T., Ehle, H. T., and Bruno, R. L. (1990). An assessment of measures to predict the outcome of driving evaluations in patients with cerebral damage. *American Journal of Occupational Therapy, 44,* 709–713.

Galski, T., Williams, J., and Ehle, H. (2000). Effect of opiods on driving ability. *Journal of Pain Symptoms Management, 19,* 200–208.

Garretson, M., and Peck, R. C. (1981). *Factors Associated with Fatal Accident Involvement Among California Drivers.* Report No. 79. Sacramento: California Department of Motor Vehicles.

Garretson, M., and Peck, R. C. (1982). Factors associated with fatal accident involvement among California drivers. *Journal of Safety Research, 13,* 141–156.

Gebers, M. A. (1998). Exploratory multivariable analyses of California driver-record accident rates. *Transportation Research Record, 1635,* 72–80.

Gebers, M. A. (1999). *Strategies for Estimating Driver Accident Risk in Relation to California's Negligent-Operator Point System.* Technical Monograph 183. Sacramento: California Department of Motor Vehicles.

Gebers, M. A. (2003). *An Inventory of California Driver Accident Risk Factors.* CAL-DMV-RSS-03-204. Sacramento: California Department of Motor Vehicles.

Gebers, M. A., and Peck, R. C. (1994). *An Inventory of California Driver Accident Risk Factors.* CAL-DMV-RSS-94-144. Sacramento: California Department of Motor Vehicles.

Gebers, M. A., and Peck, R. C. (2000). *Using Traffic Conviction Correlates To Identify High Accident-Risk Drivers.* CAL-DMV-RSS-00-187. Sacramento: California Department of Motor Vehicles.

Gebers, M. A., and Peck, R. C. (2003). Using traffic conviction correlates to identify high accident-risk drivers. *Accident Analysis and Prevention, 35,* 903–912.

Gengo, F. M., and Manning, C. (1990). A review of the effects of antihistamines on mental processes related to automobile driving. *Journal of Allergy and Clinical Immunology, 86,* 1034–1039.

Ghiselli, E. E., and Brown, C. W. (1955). *Personnel and Industrial Psychology.* Second edition. New York: McGraw-Hill Book Co Inc.

Ghosh, S. N., and Tripathi, R. S. (1965). Perceptual-motor speed ratio and accident proneness. *Indian Journal of Applied Psychology, 2,* 10–16.

Gilley, D., Wilson, R., Bennett, D., Stebbins, G., Bernard, B., Whalen, M., and Fox, J. (1991). Cessation of driving and unsafe motor vehicle operation by dementia patients. *Archives of Internal Medicine, 151,* 941–945.

Gnardellis, C., Tzamalouka, G., Papadakaki, M., and Chliaoutakis, J. E. (2008). An investigation of the effect of sleepiness, drowsy driving, and lifestyle on vehicle crashes. *Transportation Research Part F, 11,* 270–281.

Godley, S. T., Triggs, T. J., and Fildes, B. N. (2002). Driving simulator validation for speed research. *Accident Analysis and Prevention, 34,* 589–600.

Goldenbeld, C., Twisk, D., and de Craen, S. (2004). Short and long term effects of moped rider training: A field experiment. *Transportation Research Part F, 7,* 1–16.

Golding, A. P. (1983). *Differential Accident Involvement: A Literature Survey.* PERS 356:44. Council for Scientific and Industrial Research Reports.

Goldstein, L. G., and Mosel, J. N. (1958). A factor study of drivers' attitudes, with further study on driver aggression. *Highway Research Board Bulletin, 172.*

Gonzales, M. M., Dickinson, L. M., DiGuiseppi, C., and Lowenstein, S. R. (2005). Student drivers: A study of fatal motor vehicle crashes involving 16-year-old drivers. *Annals of Emergency Medicine, 45,* 140–146.

Gonzales-Rothi, R., Foresman, G., and Block, A. (1988). Do patients with sleep apnea die in their sleep? *Chest, 94,* 531–538.

Goode, K. T., Ball, K. K., Sloane, M., Roenker, D. L., Roth, D. L., Myers, R. S., and Owsley, C. (1998). Useful field of view and other neurocognitive indicators of crash risk in older adults. *Journal of Clinical Psychology in Medical Settings, 5,* 425–440.

Gordon, J. E., Gulati, P. V., and Wyon, J. B. (1962). Reliability of recall data in traumatic accidents. *Archives of Environmental Health, 4,* 575–578.

Gras, M. E., Sullman, M. J., Cunill, M., Planes, M., Aymerich, M., and Font-Mayolas, S. (2006). Spanish drivers and their aberrant driving behaviours. *Transportation Research Part F, 9,* 129–137.

Grayson, G. B., and Hakkert, A. S. (1987). Accident analysis and conflict behaviour. In J. Rothengatter and R. de Bruin (eds) *Road Users and Traffic Safety*, 27–60. Maastricht: van Gorcum.

Grayson, G. B., and Maycock, G. (1988). From proneness to liability. In J. Rothengatter and R. de Bruin (eds) *Road User Behaviour: Theory and Research*, 234–241. Maastricht: van Gorcum.

Greenshields, B. D., and Platt, F. N. (1967). Development of a method of predicting high-accident and high-violation drivers. *Journal of Applied Psychology, 51,* 205–210.

Greenwood, M., and Woods, H. M. (1919). *The Incidence of Industrial Accidents upon Individuals with Specific Reference to Multiple Accidents.* Industrial Fatigue Research Board Report No 4. London: His Majesty's Stationery Office.

Gregersen, N.-P. (1994). Systematic cooperation between driving schools and parents in driver education, an experiment. *Accident Analysis and Prevention, 26,* 453–461.

Gresset, J., and Meyer, F. (1994a). Risk of automobile accidents among elderly drivers with impairments or chronic diseases. *Canadian Journal of Public Health, 85,* 282–285.

Gresset, J., and Meyer, F. (1994b). Risk of accidents among elderly car drivers with visual acuity equal to 6/12 or 6/15 and lack of binocular vision. *Ophthalmic and Physiological Optics, 14,* 33–37.

Groeger, J. A., and Grande, G. E. (1996). Self-preserving assessments of skill? *British Journal of Psychology, 87,* 61–79.

Guibert, R., Duarte-Franco, E., Ciampi, A., Potvin, L., Lioselle, J., and Philibert, L. (1998). Medical conditions and the risk of motor vehicle crashes in men. *Archives of Family Medicine, 7,* 554–558.

Guilford, J. S. (1973). Prediction of accidents in a standardized home environment. *Journal of Applied Psychology, 57,* 306–313.

Gulian, E., Glendon, A. I., Matthews, G., Davies, D. R., and Debney, L. M. (1988). Exploration of driver stress using self-reported data. In J. Rothengatter and R. de Bruin (eds) *Road User Behaviour: Theory and Research*, 342–347. Maastricht: van Gorcum.

Gulliver, P., and Begg, D. (2007). Personality factors as predictors of persistent risky driving behavior and crash involvement among young adults. *Injury Prevention, 13,* 376–381.

Gully, S. M., Whitney, D. J., and Vanosdall, F. E. (1995). Prediction of police officers' traffic accident involvement using behavioral observations. *Accident Analysis and Prevention, 27,* 355–362.

Gumpper, D. C., and Smith, K. R. (1968). The prediction of individual accident liability with an inventory measuring risk-taking tendency. *Traffic Safety Research Review, 12,* 50–55.

Gutshall, R. W., Harper, C., and Burke, D. (1968). An exploratory study of the interrelations among driving ability, driving exposure, and socio-economic status of low, average and high intelligence males. *Exceptional Children, 35,* 43–47.

Haapanen, N., Miilunpalo, S., Pasanen, M., Oja, P., and Vuori, I. (1997). Agreement between questionnaire data and medical records of chronic diseases in middle-aged and elderly Finnish men and women. *American Journal of Epidemiology, 145,* 762–769.

Haddon, W. A., Suchman, E., and Klein, D. (1964). *Accident Research: Methods and Approaches.* New York: Harper and Row.

Haglund, M., and Åberg, L. (2000). Speed choice in relation to speed limit and influences from other drivers. *Transportation Research Part F, 3,* 39–51.

Haight, F. A. (1973). Induced exposure. *Accident Analysis and Prevention, 5,* 111–126.

Hakamies-Blomqvist, L. (1998). Older drivers' accident risk: Conceptual and methodological issues. *Accident Analysis and Prevention, 30,* 293–297.

Hakamies-Blomqvist, L., Raitanen, T., and O'Neill, D. (2002). Driver ageing does not cause higher accident rates per km. *Transportation Research Part F, 5,* 271–274.

Hakamies-Blomqvist, L., Wiklund, M., and Henriksson, P. (2005). Predicting older drivers' accident involvement – Smeed's law revisited. *Accident Analysis and Prevention, 37,* 675–680.

Hansotia, P., and Broste, S. K. (1991). The effect of epilepsy or diabetes mellitus on the risk of automobile accidents. *New England Journal of Medicine, 324,* 22–26.

Haraldsson, P. O., Carenfelt, C., Diderichsen, F., Nygren, A., and Tingvall, C. (1990). Clinical symptoms of sleep apnea syndrome and automobile accidents. *Otorhinolaryngology, 52,* 57–62.

Harano, R. M., and Peck, R. C. (1972). The effectiviness of a uniform traffic school curriculum for negligent drivers. *Accident Analysis and Prevention, 4,* 13–45.

Harano, R. M., Peck, R. C., and McBride, R. S. (1975). The prediction of accident liability through biographical data and psychometric tests. *Journal of Safety Research, 7,* 16–52.

Harlow, S. B., and Linet, M. S. (1989). Agreement between questionnaire data and medical records. The evidence for accuracy of recall. *American Journal of Epidemiology, 129,* 233–245.

Harré, N. (2003). Discrepancy between actual and estimated speeds of drivers in the presence of child pedestrians. *Injury Prevention, 9,* 38–41.

Harré, N., Brandt, T., and Dawe, M. (2000). The development of risky driving in adolescence. *Journal of Safety Research, 31,* 185–194.

Harrington, D. M. (1972). The young driver follow-up study: An evaluation of the role of human factors in the first four years of driving. *Accident Analysis and Prevention, 4,* 191–240.

Harris, S. (1990). The real number of road traffic accident casualties in the Netherlands: A year-long survey. *Accident Analysis and Prevention, 22,* 371–378.

Harrison, W. A. (2004). Investigation of the driving experience of a sample of Victorian learner drivers. *Accident Analysis and Prevention, 36,* 885–891.

Hartley, L. R., and El Hassani, J. (1994). Stress, violations and accidents. *Applied Ergonomics, 25,* 221–230.

Hartman, M. L., and Rawson, H. E. (1992). Differences in and correlates of sensation seeking in male and female athletes and nonathletes. *Personality and Individual Differences, 13,* 805–812.

Hartos, J. L., Eitel, P., and Simons-Morton, B. G. (2002). Parenting practices and adolescent risky driving: A three-month prospective study. *Health Education and Behavior, 29,* 194–206.

Haselkorn, J., Mueller, B., and Rivara, F. (1998). Characteristics of drivers and driving record after traumatic and nontraumatic brain injury. *Archives of Physical Medicine and Rehabilitation, 79,* 738–742.

Hatakka, M., Keskinen, E., Katila, A., and Laapotti, S. (1997). Self-reported driving habits are valid predictors of violations and accidents. In T. Rothengatter and E. C. Vaya, *Traffic and Transport Psychology: Theory and Application,* 295–303. Amsterdam: Pergamon.

Hauer, E. (1982). Traffic conflicts and exposure. *Accident Analysis and Prevention, 14,* 359–364.

Hauer, E., and Hakkert, A. S. (1988). Extent and some implications of incomplete accident reporting. *Transportation Research Record, 1185,* 1–10.

Hauer, E., Persaud, B. N., Smiley, A., and Duncan, D. (1991). Estimating the accident potential of an Ontario driver. *Accident Analysis and Prevention, 23,* 133–152.

Heath, E. D. (1959). Relationships between driving records, selected personality characteristics and biographical data on traffic offenders and nonoffenders. *Highway Research Board Bulletin, 212,* 16–20.

Heath, E. D. (1962). What's wrong with research in traffic accident prevention? *Traffic Digest and Review, 6,* 4–6, 14.

Hedlund, J. H. (1994). 'If they didn't drink, would they crash anyway?': The role of alcohol in traffic crashes. *Alcohol, Drugs and Driving, 10,* 115–125.

Heikkilä, V.-M., Turkka, J., Korpelainen, J., Kallanranta, T., and Summala, H. (1998). Decreased driving ability in people with Parkinson's disease. *Journal of Neurological Neurosurgical Psychiatry, 64,* 325–330.

Helander, C. J. (1984). Intervention strategies for accident-involved drivers: An experimental evaluation of current California policy and alternatives. *Journal of Safety Research, 15,* 23–40.

Hemmelgarn, B., Suissa, S., Huang, A., Boivin, J. F., and Pinard, G. (1997). Benzodiazepine use and the risk of motor vehicle crash in the elderly. *JAMA, 278,* 27–31.

Herms, B. F. (1972). Pedestrian crosswalk study: Accidents in painted and unpainted crosswalks. *Highway Research Record, 406,* 1–13.

Hingson, R., Heeren, T., Mangione, T., Morelock, S., and Mucatel, M. (1982). Teenage driving after using marijuana or drinking and traffic accident involvement. *Journal of Safety Research, 13,* 33–38.

Hjälmdahl, M., and Várhelyi, A. (2004). Validation of in-car observations, a method for driver assessment. *Transportation Research Part A, 38,* 127–142.

Holroyd, E. M. (1992). The variation of drivers' accident rates between drivers and over time. *Accident Analysis and Prevention, 24,* 275–305.

Holt, P. C. (1982). Stressful life events preceding road traffic accidents. *Injury: The British Journal of Accident Surgery, 13,* 111–115.

Hopkin, J. M., Murray, P. A., Pitcher, M., and Galasko, C. S. (1993). *Police and Hospital Recording of Non-Fatal Road Accident Casualties: A Study in Greater Manchester.* TRL Research Report 379. Crowthorne: Transport Research Laboratory.

Hormia, A. (1961). Does epilepsy mean higher susceptibility to traffic accidents? *Acta Psychiatria Neurologia Scandinavia Supplement, 150,* 210–212.

Horn, D. A. (1947). A study of pilots with repeated accidents. *Journal of Aviatic Medicine, 18,* 440–449.

Horswill, M. S., and McKenna, F. P. (1999). The development, validation, and application of a videobased technique for measuring an everyday risk-taking behavior: Drivers' speed choice. *Journal of Applied Psychology, 84,* 977–985.

Horwood, L. J., and Fergusson, D. M. (2000). Drink driving and traffic accidents in young people. *Accident Analysis and Prevention, 32,* 805–814.

Hours, M., Fort, E., Charnay, P., Bernard, M., Martin, J. L., Boisson, D., Sancho, P.-O., and Laumon, B. (2008). Diseases, consumption of medicines and responsibility for a road crash: A case-control study. *Accident Analysis and Prevention, 40,* 1789–1796.

Hu, P. S., Trumble, D. A., Foley, D. J., Eberhard, J. W., and Wallace, R. B. (1998). Crash risks of older drivers: A panel data analysis. *Accident Analysis and Prevention, 30,* 569–581.

Huebner, K. D., Porter, M. M., and Marshall, S. C. (2006). Validation of an electronic device for measuring driving exposure. *Traffic Injury Prevention,* *7,* 76–80.

Humphriss, D. (1987). Three South African studies on relation between road accidents and driver's vision. *Ophthalmic and Physiological Optics, 7,* 73–79.

Hunt, L., Morris, J. C., Edwards, D., and Wilson, B. S. (1993). Driving performance in persons with mild senile dementia of the Alzheimer type. *Journal of the American Geriatrics Society, 41,* 747–753.

Hunter, W. W., Stewart, J. R., Stutts, J. C., and Rodgman, E. A. (1993). Observed and self-reported seat belt wearing as related to prior traffic accidents and convictions. *Accident Analysis and Prevention, 25,* 545–554.

Hunter, W. W., Stutts, J. C., Stewart, J. R., and Rodgman, E. A. (1990). Characteristics of seat belt users and non-users in a state with a mandatory belt use law. *Health Education Research, 5,* 161–173.

Häkkinen, S. (1958). *Traffic Accidents and Driver Characteristics.* Unpublished doctoral thesis, Finland's Institute of Technology.

Häkkinen, S. (1979). Traffic accidents and professional driver characteristics: A follow-up study. *Accident Analysis and Prevention, 11,* 7–18.

Häkkänen, H., and Summala, H. (2000). Driver sleepiness-related problems, health status, and prolonged driving among professional Heavy-vehicle driver. *Transportation Human Factors, 2,* 151–171.

Ingham, R. (1991). The young driver. In *New Insight into Driver Behaviour.* United Kingdom: Parliamentary Advisory Council for Transport Safety.

Isherwood, J., Adam, K. S., and Hornblower, A. R. (1982). Life event stress, psychosocial factors, suicide attempts and auto-accident proclivity. *Journal of Psychosomatic Research, 26,* 371–383.

Ivan, J. N., Pasupathy, R. K., and Ossenbruggen, P. J. (1999). Differences in causality factors for single and multi-vehicle crashes on two-lane roads. *Accident Analysis and Prevention, 31,* 695–704.

Iversen, H., and Rundmo, T. (2002). Personality, risky driving and accident involvement. *Personality and Individual Differences, 33,* 1251–1263.

Iversen, H., and Rundmo, T. (2004). Attitudes towards traffic safety, driving behaviour and accident involvement among the Norwegian public. *Ergonomics,* *47,* 555–572.

James, H. F. (1991). Under-reporting of traffic accidents. *Traffic Engineering and Control, 32,* 574–584.

Jamison, K., and McGlothlin, W. H. (1973). Drug usage, personality, attitudinal, and behavioral correlates of driving behavior. *The Journal of Psychology, 83,* 123–130.

Janke, M. K. (1991). Accidents, mileage and the exaggeration of risk. *Accident Analysis and Prevention, 23,* 183–188.

Janke, M. K. (1994). Mature driver improvement program in California. *Transportation Research Record, 1438,* 77–83.

Jelalian, E., Alday, S., Spirito, A., Rasile, D., and Nobile, C. (2000). Adolescent motor vehicle crashes: The relationship between behavioural factors and self-reported injury. *Journal of Adolescent Health, 27,* 84–93.

Jerome, L., Habinski, L., and Segal, A. (2006). Attention-deficit/hyperactivity disorder (ADHD) and driving risk: A review of the literature and a methodological critique. *Current Psychiatry Report, 8,* 416–426.

Jick, H., Hunter, J. R., Dinan, B. J., Madsen, S., and Stergachis, A. (1981). Sedating drugs and automobile accidents leading to hospitalization. *American Journal of Public Health, 71,* 1399–1400.

Johns, G. (1994). How often were you absent – a review of the use of self-reported absence data. *Journal of Applied Psychology, 79,* 574–591.

Johnson, C. A., and Keltner, J. L. (1983). Incidence of visual field loss in 20,000 eyes and its relationship to driving performance. *Archives of Ophthalmology, 101,* 371–375.

Johnson, H. M. (1938). Biographical methods of detecting accident-prone drivers. *Psychological Bulletin, 35,* 511–512.

Joly, P., Joly, M.-F., Desjardins, D., Messier, S., Maag, U., Ghadirian, P., and Laberge-Nadeau, C. (1993). Exposure for different license categories through a phone survey: Validity and feasibility studies. *Accident Analysis and Prevention, 25,* 529–536.

Jonah, B. A. (1990). Age differences in risky driving. *Health Education Research, 5,* 139–149.

Jonah, B. A. (1997). Sensation seeking and risky driving: A review and synthesis of the literature. *Accident Analysis and Prevention, 29,* 651–665.

Jonah, B. A., Dawson, N. E., and Bragg, B. W. (1981). Predicting accident involvement with the Motorcycle Operator Skill Test. *Accident Analysis and Prevention, 13,* 307–318.

Junger, M., Terlouw, G.-J., and van der Heijden, P. G. (1995). Crime and accident involvement in young road users. In G. B. Grayson (ed.), *Behavioural Research in Road Safety V.* 35–54. Crowthorne: Transport Research Laboratory.

Kahneman, D., Ben-Ishai, D., and Lotan, M. (1973). Relation of a test of attention to road accidents. *Journal of Applied Psychology, 58,* 113–115.

Kahneman, D., and Tversky, A. (1972). Subjective probability: A judgement of representativiness. *Cognitive Psychology, 3,* 430–454.

Karlaftis, M. G., Kotzampassakis, I., and Kanellaidis, G. (2003). An empirical investigation of European drivers' self-assessment. *Journal of Safety Research, 34,* 207–213.

Keehn, J. D. (1959). Factor analysis of reported minor mishaps. *Journal of Applied Psychology, 43,* 311–314.

Kim, K., Li, L., Richardson, J., and Nitz, L. (1998). Drivers at fault: Influences of age, sex, and vehicle type. *Journal of Safety Research, 29,* 171–179.

King, G. F., and Clark, J. A. (1962). Perceptual-motor speed discrepancy and deviant driving. *Journal of Applied Psychology, 46,* 115–119.

Kirchner, W. K. (1961). The fallacy of accident proneness. *Personnel, 38,* 34–37.

Kishore, G. S., and Jha, S. (1978). Psycho-motor efficiency as a factor in traffic accidents.*Psychologia: An International Journal of Psychology in the Orient, 21,* 107–110.

Klen, T., and Ojanen, K. (1998). The correspondence of self-reported accidents with company records. *Safety Science, 28,* 45–48.

Kloeden, C. N., McLean, A. J., Moore, V. M., and Ponte, G. (1997). *Travelling Speed and the Risk of Crash Involvement. Volume 1: Findings.* Federal Office of Road Safety Rep. No. CR 172. Australia: Department of Transport and Regional Development. Available at http://raru.adelaide.edu.au/speed/.

Knee, C. R., Neighbors, C., and Vietor, N. A. (2001). Self-determination theory as a framework for understanding road rage. *Journal of Applied Social Psychology, 31,* 889–904.

Knopfler, M. (1982). Industrial Disease. *Love Over Gold.*

Knouse, L. E., Bagwell, C. L., Barkley, R. A., and Murphy, K. R. (2005). Accuracy of self-evaluation in adults with ADHD: Evidence from a driving study. *Journal of Attention Disorders, 8,* 221–234.

Kontogiannis, T., Kossiavelou, Z., and Marmaras, N. (2002). Self reports of aberrant behaviour on the roads: Errors and violations in a sample of Greek drivers. *Accident Analysis and Prevention, 34,* 381–399.

Koushki, P. A., and Bustan, M. (2006). Smoking, belt use, and road accidents of youth in Kuwait. *Safety Science, 44,* 733–746.

Kraft, M. A., and Forbes, T. W. (1944). Evaluating the influences of personal characteristics on the traffic accident experience of transit operators. *Proceedings of the Highway Research Board, 24,* 278–291.

Krauss, G. L., Krumholz, A., Carter, R. C., Li, G., and Kaplan, P. (1999). Risk factors for seizure-related motor vehicle crashes in patients with epilepsy. *Neurology, 52,* 1324–1329.

Kunkle, E. C. (1946). The psychological background of 'pilot error' in aircraft accidents. *Journal of Aviation Medicine, 17,* 533–567.

Köhnken, G., and Brockmann, C. (1987). Unspecific postevent information, attribution of responsibility, and eyewitness performance. *Applied Cognitive Psychology, 1,* 197–207.

Laapotti, S., Keskinen, E., Hatakka, M., and Katila, A. (2001). Novice drivers' accidents and violations – a failure on higher or lower hierarchical levels of driving behaviour. *Accident Analysis and Prevention, 33,* 759–769.

Laapotti, S., and Keskinen, E. (2004). Has the difference in accident patterns between male and female drivers changed between 1984 and 2000? *Accident Analysis and Prevention, 36,* 577–584.

Lajunen, T., Corry, A., Summala, H., and Hartley, L. (1997). Impression management and self-deception in traffic behaviour inventories. *Personality and Individual Differences 22,* 341–353.

Lajunen, T., Corry, A., Summala, H., and Hartley, L. (1998). Cross-cultural differences in drivers' self-assessments of their perceptual-motor and safety skills: Australians and Finns. *Personality and Individual Differences, 24,* 539–550.

Lajunen, T., and Parker, D. (2001). Are aggressive people aggressive drivers? A study of the relationship between self-reported general aggressiveness, driver anger and aggressive driving. *Accident Analysis and Prevention, 33,* 243–255.

Lajunen, T., and Summala, H. (1995). Driving experience, personality, and skill and safety-motive dimensions in drivers' self-assessments. *Personality and Individual Differences, 19,* 307–318.

Lajunen, T., and Summala, H. (2003). Can we trust self-reports of driving? Effects of impression management on driver behaviour questionnaire responses. *Transportation Research Part F, 6,* 97–107.

Lambert, N. M. (1995). *Analysis of Driving Histories of ADHD Subjects.* Pub. no DOT HS 808 417, 1–21. Washington: Department of Transportation, National Highway Traffic Safety Administration.

Landen, D. D., and Hendricks, S. (1995). Effect of recall on reporting at-work injuries. *Public Health Report, 110,* 350–354.

Langford, J., and Koppel, S. (2006). Epidemiology of older driver crashes – identifying older driver risk factors and exposure patterns. *Transportation Research Part F, 9,* 309–321.

Langford, J., Methorst, R., and Hakamies-Blomqvist, L. (2006). Older drivers do not have a high crash risk – a replication of low mileage bias. *Accident Analysis and Prevention, 38,* 574–578.

Langley, J. D. (1988). The need to discontinue the use of the term accident when referring to unintentional injury events. *Accident Analysis and Prevention, 20,* 1–8.

Langley, J. D., Cecchi, J. C., and Williams, S. M. (1989). Recall of injury events by thirteen year olds. *Methods of Information in Medicine, 28,* 24–27.

Lardelli-Claret, P., de Dios Luna-Del-Castillo, J., Jimenez-Moleon, J. J., Femia-Marzo, P., Moreno-Abril, O., and Bueno-Cavanillas, A. (2002). Does vehicle color influence the risk of being passively involved in a collision? *Epidemiology, 13,* 721–724.

Larson, G. E., and Merritt, C. R. (1991). Can accidents be predicted? An empirical test of the Cognitive Failures Questionnaire. *Applied Psychology: An International Review, 40,* 37–45.

Lawshe, C. H. (1940). Studies in automobile speed on the highway III. Some driver opinions and their relationship to speed on the open highway. *Journal of Applied Psychology, 24,* 318–324.

Lawton, R., and Parker, D. (1998). Individual differences in accident liability: A review and integrative approach. *Human Factors, 40,* 655–671.

Lawton, R., Parker, D., Stradling, S. G., and Manstead, A. S. (1997a). Predicting road traffic accidents: The role of social deviance and violations. *British Journal of Psychology, 88,* 249–262.

Lawton, R., Parker, D., Stradling, S. G., and Manstead, A. S. (1997b). Self-reported attitude towards speeding and its possible consequences in five different road contexts. *Journal of Community and Applied Social Psychology, 7,* 153–165.

Lee, H. C., Cameron, A. H., and Lee, A. G. (2003). Assessing the performance of older adult drivers: On-road versus simulated driving. *Accident Analysis and Prevention, 35,* 797–803.

Lee, H. C., Lee, A. H., Cameron, A. H., and Li-Tsang, C. (2003). Using a driving simulator to identify older drivers at inflated risk of motor vehicle crashes. *Journal of Safety Research, 34,* 453–459.

Lefeve, B. A., Billion, C. E., and Cross, E. C. (1956). Relation of accidents to speed habits and other driver characteristics. *Highway Research Board Bulletin, 120,* 6–30.

Legree, P. J., Heffner, T. S., Psotka, J., Martin, D. E., and Medsker, G. J. (2003). Traffic crash involvement: Experiential driving knowledge and stressful contextual antecedents. *Journal of Applied Psychology, 88,* 15–26.

Lenguerrand, E., Martin, J.-L., Moskal, A., Gadegbeku, B., and Laumon, B. (2008). Limits of the quasi-induced exposure method when compared with the standard case-control design. Application to the estimation of risks associated with driving under the influence of cannabis or alcohol. *Accident Analysis and Prevention, 40,* 861–868.

Lester, J. (1991). *Individual Differences in Accident Liability: Review of the Literature.* TRRL Research Report 306. Crowthorne: Transport and Road Research Laboratory.

Leveille, S. G., Buchner, D. M., Koepsell, T. D., McCloskey, L. W., Wolf, M. E., and Wagner, E. H. (1994). Psychoactive medications and injurious motor vehicle collisions involving older drivers. *Epidemiology, 5,* 591–598.

Levelt, P. B., and Rappange, F. (2000). Emotions and moods in car drivers and lorry drivers. Proceedings of the International Conference on Traffic and Transport Psychology. Bern: Switzerland.

Liddell, F. D. (1982). Motor vehicle accidents (1973–6) in a cohort of Montreal drivers. *Journal of Epidemiology and Community Health, 36,* 140–145.

Lie, A., Tingvall, C., Krafft, M., and Kullgren, A. (2006). The effectiviness of electronic stability control (ESC) in reducing real life crashes and injuries. *Traffic Injury Prevention, 7,* 38–43.

Lings, S. (2001). Increased driving accident frequency in Danish patients with epilepsy. *Neurology, 57,* 435–439.

Lortie, M., and Rizzo, P. (1999). The classification of accident data. *Safety Science, 31,* 31–57.

Lourens P. F., van der Molen, H. H., and Oude Egberink, H. J. (1991). Drivers and children: A matter of education? *Journal of Safety Research, 22,* 105–115.

Lourens P. F., Vissers, J. A., and Jessurun, M. (1999). Annual mileage, driving violations, and accident involvement in relation to drivers' sex, age, and level of education. *Accident Analysis and Prevention, 31,* 593–597.

Lowenstein, S. R., and Koziol-McLain, J. (2001). Drugs and traffic crash responsibility: A study of injured motorists in Colorado. *The Journal of Trauma: Injury, Infection, and Critial Care, 50,* 313–320.

Lund, A. K., and Williams, A. F. (1985). A review of the literature evaluating the defensive driving course. *Accident Analysis and Prevention, 17,* 449–460.

Lusk, S., Ronis, D., and Baer, L. (1995). A comparison of multiple indicators: Observations, supervisor report, and self-report measures of worker's hearing protection use. *Evaluation and the Health Professions, 18,* 51–63.

Maag, U., Vanasse, C., Dionne, G., and Laberge-Nadeau, C. (1997). Taxi drivers' accidents: How binocular vision problems are related to their rate and severity in terms of the number of victims. *Accident Analysis and Prevention, 29,* 217–224.

Maas, M. W., and Harris, S. (1984). Police recording of road accident in-patients: Investigation into the completeness, representativiness and reliability of police records of hospitalized traffic victims. *Accident Analysis and Prevention, 16,* 167–184.

Machin, M. A., and De Souza, J. M. (2004). Predicting health outcomes and safety behaviour in taxi drivers. *Transportation Research Part F, 7,* 257–270.

Machin, M. A., and Sankey, K. S. (2008). Relationships between young drivers' personality characteristics, risk perception, and driving behaviour. *Accident Analysis and Prevention, 40,* 541–547.

Maisto, S. A., Carter Sobell, L., Zelhart, P. F., Connors, G. J., and Cooper, T. (1979). Driving records of persons convicted of driving under the influence of alcohol. *Journal of Studies on Alcohol, 40,* 70–77.

Maki, M., and Linnoila, M. (1976). Traffic accident rates among Finnish out-patients. *Accident Analysis and Prevention, 8,* 39–44.

Malta, L. S., Blanchard, E. B., and Freidenberg, B. M. (2005). Psychiatric and behavioral problems in aggressive drivers. *Behavior Research and Therapy, 43,* 1467–1484.

Mann, H. N., and Sullman, M. J. (2008). Pre-driving attitudes and non-driving road user behaviours: Does the past predict future driving behaviour? In L. Dorn (ed.) *Driver Behaviour and Training, Volume III,* 65–73. Aldershot: Ashgate. Third International Conference on Driver Behaviour and Training. Dublin 12–13 November, 2007.

Marcotte, T. D., Lazzaretto, D., Cobb Scott, J., Roberts, E., Woods, S. P., Letendre, S., and HNRC Group. (2006). Visual attention deficits are associated with driving accidents in cognitively-impaired HIV-infected individuals. *Journal of Clinical Experimental Neuropsychology, 28,* 13–28.

Maritz, J. S. (1950). On the validity of inferences drawn from the fitting of Poisson and negative binomial distribution to observed accident data. *Psychological Bulletin, 47,* 434–443.

Markey, K. A., Buttress, S. C., and Harland, D. G. (1998). *The Characteristics and Attitudes of Adult Non-Wearers of Rear-Restraints.* TRL Report 222. Crowthorne: Transport Research Laboratory.

Marottoli, R. A., Cooney, L. M., and Tinetti, M. E. (1997). Self-report versus state records for identifying crashes among older drivers. *Journal of Gerontology Series: Biological Sciences and Medicine Sciences, 52,* 184–187.

Marottoli, R. A., Cooney, L. M., Wagner, D. R., Doucette, J., and Tinetti, M. E. (1994). Predictors of automobile crashes and moving violations among elderly drivers. *Annals of Internal Medicine, 121,* 842–846.

Marottoli, R. A., and Richardson, E. D. (1998). Confidence in, and self-rating of, driving ability among older drivers. *Accident Analysis and Prevention, 30,* 331–336.

Marottoli, R. A., Richardson, E. D., Stowe, M. H., Miller, E. G., Brass, L. M., Cooney, L. M., and Tinetti, M. E. (1998). Development of a test battery to identify older drivers at risk for self-reported adverse driving events. *Journal of the American Geriatric Society, 46,* 562–568.

Marsch, P., and Kendrick, D. (2000). Near miss and minor injury information – can it be used to plan and evaluate injury prevention programmes? *Accident Analysis and Prevention, 32,* 345–354.

Massie, D. L., Campbell, K. L., and Williams, A. F. (1995). Traffic accident involvement rates by driver age and gender. *Accident Analysis and Prevention, 27,* 73–87.

Massie, D. L., Green, P. E., and Campbell, K. L. (1997). Crash involvement rates by driver gender and the role of average annual mileage. *Accident Analysis and Prevention, 29,* 675–685.

Matthews, G., Desmond, P. A., Joyner, L., Carcary, B., and Gilliland, K. (1996). *Validation of the Driver Stress Inventory and Driver Coping Questionnaire.* Unpublished report.

Matthews, G., Desmond, P. A., Joyner, L., Carcary, B., and Gilliland, K. (1997). A comprehensive questionnaire measure of driver stress and affect. In T. Rothengatter, and E. C. Vaya (eds), *Traffic and Transport Psychology: Theory and Application,* 317–324. Amsterdam: Pergamon.

Matthews, G., Dorn, L., and Glendon, A. I. (1991). Personality correlates of driver stress. *Personality and Individual Differences, 12,* 535–549.

Matthews, G., Dorn, L., Hoyes, T. W., Davies, D. R., Glendon, A. I., and Taylor, R. G. (1998). Driver stress and performance on a driving simulator. *Human Factors, 40,* 136–149.

Matthews, G., Tsuda, A., Xin, G., and Ozeki, Y. (1999). Individual differences in driver stress vulnerability in a Japanese sample. *Ergonomics, 42,* 401–415.

Maxwell, J. P., Grant, S., and Lipkin, S. (2005). Further validation of the propensity for angry driving scale in British drivers. *Personality and Individual Differences, 38,* 213–224.

Maycock, G. (1985). Accident liability and human factors-researching the relationship. *Traffic Engineering and Control, 26,* 330–335.

Maycock, G. (1997). Sleepiness and driving: The experience of heavy goods vehicle drivers in the UK. *Journal of Sleep Research, 6,* 238–244.

Maycock, G., and Lester, J. (1995). Accident liability of car drivers: Follow-up study. In G. B. Grayson (ed.) *Behavioural Research in Road Safety V,* 106–120. Crowthorne: Transport Research Laboratory.

Maycock, G., and Lockwood, C. R. (1993). The accident liability of British car drivers. *Transport Reviews, 13,* 213–245.

Maycock, J., Lockwood, C., and Lester, J. F. (1991). *The Accident Liability of Car Drivers.* TRRL Research Report No. 315. Crowthorne: Transport and Road Research Laboratory.

Mayhew, D. R., Simpson, H. M., and Pak, A. (2003). Changes in collision rates among novice drivers during the first months of driving. *Accident Analysis and Prevention, 35,* 683–691.

McBain, W. N. (1970). Arousal, monotony, and accidents in line driving. *Journal of Applied Psychology, 54,* 509–519.

McBride, R. S. (1970). Prediction of driving behavior following a group driver improvement session. *Journal of Applied Psychology, 54,* 45–50.

McCartt, A. T., Shabanova, V. I., and Leaf, W. A. (2003). Driving experience, crashes and traffic citations of teenage beginning drivers. *Accident Analysis and Prevention, 35,* 311–320.

McDavid, J. C., Lohrmann, B. A., and Lohrmann, G. (1989). Does motorcycle training reduce accidents? Evidence from a longitudinal quasi-experimental study. *Journal of Safety Research, 20,* 61–72.

McDonald, L., Parker, D., Sutcliffe, P., and Rabbitt, P. (2000). Testing older drivers on the road. *Behavioural Research in Traffic Safety X,* 163–180. London: Department for Transport.

McEvoy, S. P., Stevenson, M. R., and Woodward, M. (2006). Phone use and crashes while driving: A representative survey of drivers in two Australian states. *Medical Journal of Australia, 185,* 630–634.

McFarland, R. A., and Moore, R. C. (1957). Human factors in highway safety: A review and evaluation. *New England Journal of Medicine, 256,* 792–799, 837–845.

McFarland, R. A., and Moore, R. C. (1962). Accidents and accident prevention. *Annual Review of Medicine, 13,* 371–388.

McGlade, F., and Laws, F. D. (1962). Classifying accidents: A theoretical viewpoint. *Traffic Safety Research Review, 6,* 2–8.

McGuire, F. L. (1956a). Rosenzweig picture frustration study for selecting safe drivers. *U.S. Armed Forces Medical Journal, 7,* 200–207.

McGuire, F. L. (1956b). The safe-driver inventory. *U.S. Armed Forces Medical Journal, 7,* 1249–1264.

McGuire, F. L. (1956c). Psychological comparison of automobile drivers. *U.S. Armed Forces Medical Journal, 7,* 1741–1748.

McGuire, F. L. (1972). Smoking, driver education, and other correlates of accidents among young males. *Journal of Safety Research, 4,* 5–11.

McGuire, F. L. (1973). The nature of bias in official accident and violation records. *Journal of Applied Psychology, 57,* 300–305.

McGuire, F. L. (1976). The validity of accident and violation criteria in the study of drinking drivers. *Journal of Safety Research, 8,* 46–47.

McGuire, F. L., and Kersch, R. C. (1969). *A Study of History, Philosophy, Research Methodology, and Effectiveness in the Field of Driver Education.* University of California Publications in Education, 19. Berkely: University of California Press.

McGwin, G., Owsley, C., and Ball, K. (1998). Identifying crash involvement among older drivers: Agreement between self-report and state records. *Accident Analysis and Prevention, 30,* 781–791.

McGwin, G., Sims, R. V., Pulley, L., and Roseman, J. M. (2000). Relations among chronic medical conditions, medications, and automobile crashes in the elderly: A population-based case-control study. *American Journal of Epidemiology, 152,* 424–431.

McKenna, F. P. (1983). Accident proneness: A conceptual analysis. *Accident Analysis and Prevention, 15,* 65–71.

McKenna, F. P., Duncan, J., and Brown, I. D. (1986). Cognitive abilities and safety on the road: A re-examination of individual differences in dichotic listening and search for embedded figures. *Ergonomics, 29,* 649–663.

McKenna, F. P., and Horswill, M. S. (2006). Risk taking from the participant's perspective: The case of driving and accident risk. *Health Psychology, 25,* 163–170.

McKenna, F. P., Horswill, M. S., and Alexander, J. L. (2006). Does anticipation training affect drivers' risk taking? *Journal of Experimental Psychology, 12,* 1–10.

McKnight, A. J., and Edwards, R. (1982). An experimental evaluation of driver license manuals and written tests. *Accident Analysis and Prevention, 14,* 187–192.

McKnight, A. J., and McKnight, A. S. (1999). Multivariate analysis of age-related driver ability and performance deficits. *Accident Analysis and Prevention, 31,* 445–454.

McKnight, A. J., and McKnight, A. S. (2003). Young novice drivers: Careless or clueless? *Accident Analysis and Prevention, 35,* 921–925.

McKnight, A. J., Shinar, D., and Hilburn, B. (1991). The visual and driving performance of monocular and binocular heavy-duty truck drivers. *Accident Analysis and Prevention, 23,* 225–237.

McMillen, D. L., Smith, S. M., and Wells-Parker, E. (1989). The effects of alcohol, expectancy, and sensation seeking on driving risk taking. *Addictive Behaviors, 14,* 477–483.

McMurray, L. (1970). Emotional stress and driving performance: The effect of divorce. *Behavioral Research in Highway Safety, 1,* 100–114.

Meadows, M. L., Stradling, S. G., and Lawson, S. (1998). The role of social deviance and violations in predicting road traffic accidents in a sample of young offenders. *British Journal of Psychology, 89,* 417–431.

Mesken, J., Lajunen, T., and Summala, H. (2002). Interpersonal violations, speeding violations and their relation to accident involvement in Finland. *Ergonomics, 45,* 469–483.

Miller, T. M., and Schuster, D. H. (1983). Long-term predictability of driver behaviour. *Accident Analysis and Prevention, 15,* 11–22.

Milosevic, S., and Vucinic, S. (1975). Statistical study of tram drivers' accidents. *Accident Analysis and Prevention, 7,* 1–7.

Mintz, A. (1954a). The inference of accident liability from the accident record. *Journal of Applied Psychology, 38,* 41–46.

Mintz, A. (1954b). Time intervals between accidents. *Journal of Applied Psychology, 38,* 401–406.

Mintz, A., and Blum, M. L. (1949). A re-examination of the accident proneness concept. *Journal of Applied Psychology, 33,* 195–211.

Moffie, D. J., and Alexander, C. (1953). Relationship of preventable to non-preventable accidents in the trucking industry. *Bulletin of the Highway Research Board, 73,* 32–41.

Moffie, D., Symmes, A., and Milton, C. (1952). Relation between psychological tests and driver performance. *Highway Research Board Bulletin, 60,* 17–24.

Mohr, D. L., and Clemmer, D. I. (1988). The 'accident prone' worker: An example from heavy industry. *Accident Analysis and Prevention, 20,* 123–127.

Moore, M., and Dahlen, E. R. (2008). Forgiveness and consideration of future consequences in aggressive driving. *Accident Analysis and Prevention, 40,* 1661–1666.

Morrow, P. C., and Crum, M. R. (2004). Antecedents of fatigue, close calls, and crashes among commercial motor-vehicle drivers. *Journal of Safety Research, 35,* 59–69.

Moshiro, C., Heuch, I., Åström, A. N., Setel, P., and Kvåle, G. (2005). Effect of recall on estimation of non-fatal injury rates: A community based study in Tanzania. *Injury Prevention, 11,* 48–52.

Moskowitz, H. (1985). Marijuana and driving. *Accident Analysis and Prevention, 17,* 323–345.

Mounce, N. H., and Pendleton, O. J. (1992). The relationship between blood alcohol concentration and crash responsibility for fatally injured drivers. *Accident Analysis and Prevention, 24,* 201–210.

Mozdzierz, G. J., Macchitelli, F. J., Planek, T. W., and Lottman, T. J. (1975). Personality and temperament differences between alcoholics with high and low records of traffic accidents and violations. *Journal of Studies on Alcohol, 36,* 395–399.

Munden, J. W. (1967). *The Relation Between a Driver's Speed and His Accident Rate.* RRL Report 88. Crowthorne: Road Research Laboratory.

Murphy, K. R., and Barkley, R. A. (1996). Prevalence of ADHD and ODD symptoms in a community sample of adult licensed drivers. *Journal of Attention Disorders, 1*, 147–161.

Murray, W., Cuerden, A., and Darby, P. (2005). Comparing IT-based driver assessment results against self-reported and actual crash outcomes in a large motor vehicle fleet. In L. Dorn (ed.), *Driver Behaviour and Training, Volume II*, 373–382. Second International Conference on Driver Behaviour and Training. Edinburgh 15–17 November, 2005. Aldershot: Ashgate.

Nabi, H., Consoli, S. M., Chastang, J. F., Chiron, M., Lafont, S., and Lagarde, E. (2005). Type A behavior pattern, risky driving behaviors, and serious road traffic accidents: A prospective study of the GAZEL cohort. *American Journal of Epidemiology, 161*, 864–870.

Nada-Raja, S., Langley, J. D., McGee, R., Williams, S. M., Begg, D. J., and Reeder, A. I. (1997). Inattentive and hyperactive behaviors and driving offences in adolescence. *Journal of the American Academy of Child and Adolescent Psychiatry, 36*, 515–522.

Nagatsuka, Y. (1970). Discriminative Reaction Test of Multiple Performance Type as applied to hire-taxi drivers and truck drivers. *Tohoku Psychologica Folia, 23*, 116–121.

Nasvadi, G. E., and Vavrik, J. (2007). Crash risk of older drivers after attending a mature driver education program. *Accident Analysis and Prevention, 39*, 1073–1079.

Newbold, E. M. (1926). *Contribution to the Study of the Human Factor in the Causation of Accidents.* Industrial Health Research Board Report No 34. London: His Majesty's Stationary Office.

Newbold, E. M. (1927). Practical applications of the statistics of repeated events, particularly industrial accidents. *Journal of the Royal Statistical Society, 90*, 487–547.

Newnam, S., Griffin, M. A., and Mason, C. (2008). Safety in work vehicles: A multilevel study linking safety values and individual predictors to work-related driving crashes. *Journal of Applied Psychology, 93*, 632–644.

Newnam, S., Watson, B., and Murray, W. (2004). Factors predicting intentions to speed in a work and personal vehicle. *Transportation Research Part F, 7*, 287–300.

van Nooten, W. N., Blom, D. H., Pokorny, M. L., and van Leeuwen, P. (1991). Time intervals between bus drivers' accidents. *Journal of Safety Research, 22*, 41–47.

Norris, F. H., Matthews, B. A., and Riad, J. K. (2000). Characterological, situational, and behavioral risk factors for motor vehicle accidents: A prospective examination. *Accident Analysis and Prevention, 32*, 505–515.

Norrish, A., North, D., Kirkman, P., and Jackson, R. (1994). Validity of self-reported hospital admission in a prospective study. *American Journal of Epidemiology, 140*, 938–42.

Norton, R., Vander Hoorn, S., Roberts, I., Jackson, R., and MacMahon, S. (1997). Migraine: A risk factor for motor vehicle driver injury. *Accident Analysis and Prevention, 29,* 699–701.

Osgood, C. E., and Tannenbaum, P. H. (1955). The principle of congruity in the prediction of attitude change. *Psychological Review, 62,* 42–55.

Owsley, C., Ball, K., McGwin, G., Sloane, M. E., Roenker, D. L., White, M. F., and Overley, E. T. (1998). Visual processing impairment and risk of motor vehicle crash among older adults. *JAMA, 279,* 1083–1088.

Owsley, C., Ball, K., Sloane, M. E., Roenker, D. L., and Bruni, J. R. (1991). Visual/cognitive correlates of vehicle accidents in older drivers. *Psychology and Aging, 6,* 403–415.

Owsley, C., McGwin, G. Jr., Phillips, J. M., McNeal, S. F., and Stalvey, B. T. (2004). Impact of an educational program on the safety of high-risk, visually impaired, older drivers. *American Journal of Preventive Medicine, 26,* 222–229.

Owsley, C., Stalvey, B., Wells, J., and Sloane, M. (1999). Older drivers and cataract: Driving habits and crash risk. *Journal of Gerontology Medical Sciences, 54A,* M203-M211.

Panek, P. E., and Rearden, J. J. (1987). Age and gender effects on accident types for rural drivers. *The Journal of Applied Gerontology, 6,* 332–346.

Panek, P. E., Wagner, E. E., Barrett, G. V., and Alexander, R. A. (1978). Selected Hand test personality variables related to accidents in female drivers. *Journal of Personality Assessment, 42,* 355–357.

Parada, M. A., Cohn, L. D., Gonzalez, E., Byrd, T., and Cortes, M. (2001). The validity of self-reported seatbelt use: Hispanic and non-Hispanic drivers in El Paso. *Accident Analysis and Prevention, 33,* 139–143.

Paris, H., and Van den Broucke, S. (2008). Measuring cognitive determinants of speeding: An application of the theory of planned behaviour. *Transportation Research Part F, 11,* 168–180.

Parker, D. (1999). Elderly drivers and their accidents: The Ageing Driver Questionnaire. In G. Grayson (ed.) *Behavioural Research in Road Safety IX,* 169–178. Crowthorne: Transport Research Laboratory.

Parker, D., Manstead, A. S., and Stradling, S. G. (1995). Extending the theory of planned behaviour: The role of personal norm. *British Journal of Social Psychology, 43,* 127–137.

Parker, D., Manstead, A. S., Stradling, S. G., and Reason, J. T. (1992). Determinants of intention to commit driving violations. *Accident Analysis and Prevention, 24,* 117–131.

Parker, D., McDonald, L., Rabbitt, P., and Sutcliffe, P. (2000). Elderly drivers and their accidents: The Aging Driver Questionnaire. *Accident Analysis and Prevention, 32,* 751–759.

Parker, D., Reason, J. T., Manstead, A. S., and Stradling, S. G. (1995). Driving errors, driving violations and accident involvement. *Ergonomics, 38,* 1036–1048.

Parker, D., Stradling, S. G., and Manstead, A. S. (1996). Modifying beliefs and attitudes to exceeding the speed limit: An intervention study based on the Theory of Planned Behavior. *Journal of Applied Psychology, 26,* 1–19.

Parker, D., West, R., Stradling, S., and Manstead, A. S. (1995). Behavioural characteristics and involvement in different types of traffic accident. *Accident Analysis and Prevention, 27,* 571–581.

Parker, J. W. (1953). Psychological and personal history data related to accident records of commercial truck drivers. *Journal of Applied Psychology, 37,* 317–320.

Parmentier, G., Chastang, J. F., Nabi, H., Chiron, M., Lafont, S., and Lagarde, E. (2005). Road mobility and the risk of road traffic accident as a driver. The impact of medical conditions and life events. *Accident Analysis and Prevention, 37,* 1121–1134.

Paulhus, D. L. (1991). Measurement and control of response bias. In J. Robinson, P. Shaver and L. Wrightsman (eds), *Measures of Personality and Social Psychological Attitudes,* 17–59. San Diego: Academic Press.

Payne, C. E., and Selzer, M. L. (1962). Traffic accidents, personality and alcoholism; a preliminary study. *Journal of Abdominal Surgery, 4,* 21–26.

Peck, R. C. (1993). The identification of multiple accident correlates in high risk drivers with specific emphasis on the role of age, experience and prior traffic violation frequency. *Alcohol, Drugs and Driving, 9,* 145–166.

Peck, R. C., and Gebers, M. A. (1992). *The California Driver Record Study: A Multiple Regression Analysis of Driver Record Histories from 1969 through 1982.* Sacramento: California Department of Motor Vehicles.

Peck, R. C., and Kuan, J. A. (1983). A statistical model of individual accident risk prediction using driver record, territory and other biographical factors. *Accident Analysis and Prevention, 15,* 371–393.

Peck, R. C., McBride, R. S., and Coppin, R. S. (1971). The distribution and prediction of driver accident frequencies. *Accident Analysis and Prevention, 2,* 243–299.

Pelz, D. C., and Krupat, E. (1974). Caution profile and driving record of undergraduate males. *Accident Analysis and Prevention, 6,* 45–58.

Pelz, D. C., and Schuman, S. H. (1968). Dangerous young drivers. *Highway Research News, 33,* 31–41.

Pérez-Chada, D., Videla, A. J., O'Flaherty, M. E., Palermo, P., Meoni, J., Sarchi, M. I., Khoury, M., and Durán-Cantolla, J. (2005). Sleep habits and accident risk among truck drivers: A cross-sectional study in Argentina. *Sleep, 28,* 1103–1108.

Perry, A. R. (1986). Type A behaviour pattern and motor vehicle drivers' behavior. *Perceptual and Motor Skills, 63,* 875–878.

Philip, P. (2005). Sleepiness of occupational drivers. *Industrial Health, 43,* 30–33.

de Pinho, R. S., da Silva-Júnior, F. P., Bastos, J. P., Maia, W. S., de Mello, M. T., de Bruin, V. M., and de Bruin, P. F. (2006). Hypersomnolence and accidents in truck drivers: A cross-sectional study. *Chronobiology International, 23,* 963–971.

Pizza, F., Contardi, S., Ferlisi, M., Mondini, S., and Cirignotta, F. (2008). Daytime driving simulation performance and sleepiness in obstructive sleep apnoea patients. *Accident Analysis and Prevention, 40,* 602–609.

Planek, T. W., Schupack, S. A., and Fowler, R. C. (1974). An evaluation of the national safety council's defensive driving course in various states. *Accident Analysis and Prevention, 6,* 271–297.

Plummer, L. S., and Das, S. S. (1973). A study of dichotomous thought processes in accident-prone drivers. *British Journal of Psychiatry, 122,* 289–294.

Podsakoff, P. M., Mackenzie, S. B., Lee, J. Y., and Podsakoff, N. P. (2003). Common method biases in behavioral research: A critical review of the literature and recommended remedies. *Journal of Applied Psychology, 88,* 879–903.

Podsakoff, P. M., and Organ, D. W. (1986). Self-reports in organizational research: Problems and prospects. *Journal of Management, 12,* 531–544.

Popham, R. E., and Schmidt, W. (1981). Words and deeds: The validity of self-report data on alcohol consumption. *Journal of Studies on Alcohol, 42,* 355–358.

Popper, K. R. (1968). *The Logic of Scientific Discovery.* Second edition. London: Hutchinson.

Porter, C. S. (1988). Accident proneness: A review of the concept. In D. J. Oborne (ed.) *International Reviews of Ergonomics: Current Trends in Human Factors Research and Practices, Vol 2,* 177–206. London: Taylor and Francis.

Porter, C. S., and Corlett, E. N. (1989). Performance differences of individuals classified by questionnaire as accident prone or non-accident prone. *Ergonomics, 32,* 317–333.

Prabhakar, T., Lee, S. H., and Job, R. F. (1996). Risk taking, optimism bias and risk utility in young drivers. In *Proceedings of the Road Safety Research and Enforcement Conference,* 61–68. Sydney: Roads and Traffic Authority of NSW.

Preusser, D. F., Williams, A. F., and Ulmer, R. G. (1995). Analysis of fatal motorcycle crashes: Crash typing. *Accident Analysis and Prevention, 27,* 845–851.

Pöysti, L., Rajalin, S., and Summala, H. (2005). Factors influencing the use of cellular (mobile) phone during driving and hazards while using it. *Accident Analysis and Prevention, 37,* 47–51.

Quenault, S. W. (1967). *Driver Behaviour – Safe and Unsafe Drivers.* RRL Report LR 70. Crowthorne: Road Research Laboratory.

Quenault, S. W. (1968). *Driver Behaviour – Safe and Unsafe Drivers. Part II.* RRL Report LR 146. Crowthorne: Road Research Laboratory.

Quimby, A. R., Maycock, G., Carter, I. D., Dixon, R., and Wall, J. G. (1986). *Perceptual Abilities of Accident Involved Drivers.* TRRL Research Report 27. Crowthorne: Transport and Road Research Laboratory.

Quimby, A., Maycock, G., Palmer, C., and Grayson, G. B. (1999). *Drivers' Speed Choice: An In-Depth Study.* TRL Report 326. Crowthorne: Transport Research Laboratory.

Quimby, A. R., and Watts, G. R. (1981). *Human Factors and Driving Performance.* TRRL Laboratory Report 1004. Crowthorne: Transport and Road Research Laboratory.

Rajalin, S. (1994). The connection between risky driving and involvement in fatal accidents. *Accident Analysis and Prevention, 26,* 555–562.

Rajalin, S., and Summala, H. (1997). What surviving drivers learn from a fatal accident. *Accident Analysis and Prevention, 29,* 277–283.

Rawson, A. J. (1944). Accident proneness. *Psychosomatic Medicine, 6,* 88–94.

Reimer, B., D'Ambrosio, L. A., Coughlin, J. F., Fried, R., and Biederman, J. (2007). Task-induced fatigue and collisions in adult drivers with Attention Deficit Hyperactivity Disorder. *Traffic Injury Prevention, 8,* 290–299.

Retting, R. A., Weinstein, H. B., Williams, A. F., and Preusser, D. F. (2001). A simple method for identifying and correcting crash problems on urban arterial streets. *Accident Analysis and Prevention, 33,* 723–734.

Retting, R. A., Williams, A. F., Preusser, D. F., and Weinstein, H. B. (1995). Classifying urban crashes for countermeasure development. *Accident Analysis and Prevention, 27,* 283–294.

Richards, T. L., Deffenbacher, J. L., and Rosén, L. A. (2002). Driving anger and other driving-related behaviors in high and low ADHD symptom college students. *Journal of Attention Disorders, 6,* 25–38.

Rimmö, P.-A. (1999). *Modelling Self-Reported Aberrant Driving Behaviour.* Unpublished doctoral thesis, Uppsala University.

Rimmö, P.-A., and Åberg, L. (1999). On the distinction between violations and errors: Sensation seeking associations. *Transportation Research Part F, 2,* 151–166.

Rimmö, P-. A., and Hakamies-Blomqvist, L. (2002). Older drivers' aberrant driving behaviour, impaired activity, and health as reasons for self-imposed driving limitations. *Transportation Research Part F, 5,* 47–62.

Risk, A., and Shaoul, J. E. (1982). Exposure to risk and the risk of exposure. *Accident Analysis and Prevention, 14,* 353–357.

Risser, R. (1985). Behavior in traffic conflict situations. *Accident Analysis and Prevention, 17,* 179–197.

Risser, R., Chaloupka, C., Grundler, W., Sommer, M., Häusler, J., and Kaufmann, C. (2008). Using non-linear methods to investigate the criterion validity of traffic-psychological test batteries. *Accident Analysis and Prevention, 40,* 149–157.

Roberts, K. L., Chapman, P. R., and Underwood, G. (2004). The relationship between accidents and near-accidents in a sample of company vehicle drivers. In T. Rothengatter and R. Huguenin (eds), *Traffic and Transport Psychology: Theory and Application*, 209–218. Amsterdam: Elsevier.

Roberts, S. E., Vingilis, E., Wilk, P., and Seely, J. (2008). A comparison of self-reported motor vehicle collision injuries compared with official collision data: An analysis of age and sex trends using the Canadian National Population Health Survey and Transport Canada data. *Accident Analysis and Prevention, 40*, 559–566.

Robertson, A. S., Rivara, F. P., Ebel, B. E., Lymp, J. F., and Christakais, D. A. (2005). Validation of parent self reported home safety practices. *Injury Prevention, 11*, 209–212.

Robertson, L. S. (1992). The validity of self-reported behavioral risk factors: Seatbelt and alcohol use. *The Journal of Trauma, 32*, 58–59.

Robertson, M. D., and Drummer, O. H. (1994). Responsibility analysis: A methodology to study the effects of drugs in driving. *Accident Analysis and Prevention, 26*, 243–247.

Rogé, J., Pébayle, T., Campagne, A., and Muzet, A. (2005). Useful visual field reduction as a function of age and risk of accident in simulated car driving. *Investigative Ophthalmology and Visusal Science, 46*, 1774–1779.

Rolls, G. W., Hall, R. D., Ingham, R., and McDonald, M. (1991). *Accident Risk and Behavioural Patterns of Younger Drivers*. AA Foundation for Road Safety Research.

Rosenbloom, T., Shahar, A., Elharar, A., and Danino, O. (2008). Risk perception of driving as a function of advanced training aimed at recognizing and handling risks in demanding driving situations. *Accident Analysis and Prevention, 40*, 697–703.

Rosman, D. L. (2001). The Western Australian Road Injury Database (1987–1996): Ten years of linked police, hospital and death records of road crashes and injuries. *Accident Analysis and Prevention, 33*, 81–88.

Rosman, D. L., and Knuiman, M. W. (1994). A comparison of hospital and police road injury data. *Accident Analysis and Prevention, 26*, 215–222.

Ross, H. L. (1966). Driving records of accident-involved drivers. *Traffic Safety Research Review, 10*, 22–25.

Rotter, J. B. (1966). Generalized expectations for internal versus external control of reinforcement. *Psychological Monographs, 80*, 609.

Rushton, J. P., Brainerd, C. J., and Pressley, M. (1983). Behavioral development and construct validity: The principle of aggregation. *Psychological Bulletin, 94*, 18–38.

Sagberg, F. (2006). Driver health and crash involvement: A case-control study. *Accident Analysis and Prevention, 38*, 28–34.

Salgado, J. F. (2002). The Big Five personality dimensions and counterproductive behaviors. *International Journal of Selection and Assessment, 10*, 117–125.

Salminen, S. (2005). Relationships between injuries at work and during leisure time. *Accident Analysis and Prevention, 37,* 373–376.

Salminen, S., and Heiskanen, M. (1997). Correlations between traffic, occupational, sports, and home accidents. *Accident Analysis and Prevention, 29,* 33–36.

Santos, J., Merat, N., Mouta, S., Brookhuis, K., and de Waard, D. (2005). The interaction between driving and in-vehicle information systems: Comparisons of results from laboratory, simulator and real-world studies. *Transportation Research Part F, 8,* 135–146.

Sass, R., and Crook, G. (1981). Accident proneness: Science or non-science? *International Journal of Health Services, 11,* 175–190.

Schmitt, N. (1994). Method bias: The importance of theory and measurement. *Journal of Organizational Behavior, 15,* 393–398.

Schuman, S. H., Pelz, D. C., Ehrlich, N. J., and Selzer, N. L. (1967). Young male drivers: Impulse expression, accidents and violations. *Journal of the American Medical Association, 200,* 1026–1036.

Schuster, D. H. (1968). Prediction of follow-up driving accidents and violations. *Traffic Safety Research Review, 12,* 17–21.

Schuster, D. H., and Guilford, J. P. (1962). The psychometric prediction of problem drivers. *Traffic Safety Research Review, 6,* 16–20.

Selzer, M. L., Rogers, J. E., and Kern, S. (1968). Fatal accidents: The role of psychopathology, social stress, and acute disturbance. *Journal of Psychiatry, 124,* 1028–1036.

Selzer, M. L., and Vinokur, A. (1975). Role of life events in accident causation. *Mental Health, 2,* 36–54.

Shaw, L. (1965). The practical use of projective personality tests as accident predictors. *Traffic Safety Research Review, 9,* 34–72.

Shaw, L., and Sichel, H. S. (1961). The reduction of accidents in a transport company by the determination of the accident liability of individual drivers. *Traffic Safety Research Review, 5,* 2–13.

Shinar, D., Treat, J. R., and McDonald, S. T. (1983). The validity of police reported data. *Accident Analysis and Prevention, 15,* 175–191.

Shope, J. T., Waller, P. F., Raghunathan, T. E., and Patil, S. M. (2001). Adolescent antecedents of high-risk driving behavior into young adulthood: Substance use and parental influence. *Accident Analysis and Prevention, 33,* 649–658.

Sichel, H. S. (1965). The statistical estimation of individual accident liability. *Traffic Safety Research Review, 9,* 8–15.

Signori, E. I., and Bowman, R. G. (1974). On the study of personality factors in research on driving behavior. *Perceptual and Motor Skills, 38,* 1067–1076.

Simon, F., and Corbett, C. (1996). Road traffic offending, stress, age, and accident history among male and female drivers. *Ergonomics, 39,* 757–780.

Simons-Morton, B. G., Hartos, J. L, Leaf, W. A., and Preusser, D. F. (2006). The effects of the Checkpoints program on parent-imposed driving limits and crash outcomes among Connecticut novice teen drivers at six-month post-licensure. *Journal of Safety Research, 37,* 9–15.

Simons-Morton, B. G., Lerner, N., and Singer, J. (2005). The observed effects of teenage passengers on the risky driving behavior of teenage drivers. *Accident Analysis and Prevention, 37,* 973–982.

Slocombe, C. S., and Brakeman, E. E. (1930). Psychological tests and accident proneness. *British Journal of Psychology, 21,* 29–38.

Smart, R. G., and Schmidt, W. S. (1962). Psychosomatic disorders and traffic accidents. *Journal of Psychosomatic Research, 6,* 191–197.

Smart, R. G., and Schmidt, W. (1969). Physiological impairment and personality factors in traffic accidents of alcoholics. *Quarterly Journal of Studies on Alcohol, 30,* 440–445.

Smeed, R. J. (1960). Proneness of drivers to road accidents. *Nature, 186,* 273–275.

Smink, B. E., Ruiter, B., Lusthof, K. J., de Gier, J. J., Uges, D. R., and Egberts, A. C. (2005). Drug use and the severity of a traffic accident. *Accident Analysis and Prevention, 37,* 427–433.

Smith, C. S., Silverman, G. S., Heckert, T. M., Brodke, M., Hayes, B., and Mattimore, L. K. (2001). A comprehensive method for the assessment of industrial injury events. *Journal of Prevention and Intervention in the Community, 22,* 5–20.

Smith, D. I. (1976). Official driver records and self-reports as sources of accident and conviction data for research purposes. *Accident Analysis and Prevention, 14,* 439–442.

Smith, D. I., and Kirkham, R. W. (1982). Relationship between intelligence and driving record. *Accident Analysis and Prevention, 14,* 439–442.

Smith, D. L., and Heckert, T. M. (1998). Personality characteristics and traffic accidents of college students. *Journal of Safety Research, 29,* 163–169.

Smith, R. N. (1966). The reporting level of California State Highway accidents. *Traffic Engineering, 29,* 20–25.

Sobel, R., and Underhill, R. (1976). Family disorganization and teenage auto accidents. *Journal of Safety Research, 8,* 8–18.

Soderstrom, C. A., Dischinger, P. C., Ho, S. M., and Soderstrom, M. T. (1993). Alcohol use, driving records, and crash culpability among injured motorcycle drivers. *Accident Analysis and Prevention, 25,* 711–716.

Sohn, S. Y., and Shin, H. (2001). Pattern recognition for road traffic accident severity in Korea. *Ergonomics, 44,* 107–117.

Sommer, M., Herle, M., Häusler, J., Risser, R., Schützhofer, B., and Chaloupka, C. (2008). Cognitive and personality determinants of fitness to drive. *Transportation Research Part F, 11,* 362–375.

Spector, P. E. (1994). Using self-report questionnaires in OB research: A comment on the use of a controversial method. *Journal of Organizational Behavior, 15,* 285–392.

Staplin, L., Gish, K. W., and Joyce, J. (2008). 'Low mileage bias' and related policy implications – a cautionary note. *Accident Analysis and Prevention, 40,* 1249–1252.

Stein, H. S., and Jones, I. S. (1988). Crash-involvement of large trucks by configuration: A case-control study. *American Journal of Public Health, 78,* 491–498.

Stewart, R. G. (1957). Reported driving speed and previous accidents. *Journal of Applied Psychology, 41,* 293–296.

Stewart, R. J., and Campbell, B. J. (1972). *The Statistical Association Between Past and Future Accidents and Violations.* Chapel Hill: University of North Carolina.

Stoohs, R. A., Guilleminault, C., Itoi, A., and Dement, W. C. (1994). Traffic accidents in commercial long-haul truck drivers: The influence of sleep-disordered breathing. *Sleep, 17,* 619–623.

Stradling, S. G., Parker, D., Lajunen, T., Meadows, M. L., and Xie, C. Q. (1998). Drivers' violations, erros, lapses and crash involvement: International comparisons. Paper presented at the 9th International Conference on Road Safety in Europe.

Streff, F. M., and Wagenaar, A. C. (1989). Are there really shortcuts? Estimating seat belt use with self-report measures. *Accident Analysis and Prevention, 21,* 509–516.

Struckman-Johnson, D. L., Lund, A. K., Williams, A. F., and Osborne, D. W. (1989). Comparative effects of driver improvement programmes on crashes and violations. *Accident Analysis and Prevention, 21,* 203–215.

Stulginskas, J. V., Verreault, R., and Pless, I. B. (1985). A comparison of observed and reported restraint use by children and adults. *Accident Analysis and Prevention, 17,* 381–386.

Stutts, J. C., Stewart, J. R., and Martell, C. (1998). Cognitive test performance and crash risk in an older driver population. *Analysis and Prevention, 30,* 337–346.

Suchman, E. A. (1970). Accidents and social deviance. *Journal of Health and Social Behavior, 11,* 4–15.

Sullman, M. J., and Baas, P. H. (2004). Mobile phone use amongst New Zealand drivers. *Transportation Research Part F, 7,* 95–105.

Sullman, M. J., Meadows, M. L., and Pajo, K. B. (2002). Aberrant driving behaviours amongst New Zealand truck drivers. *Transportation Research Part F, 5,* 217–232.

Sullman, M. J., Meadows, M. L., and Pajo, K. B. (2004). Errors, lapses and violations in the drivers of heavy vehicles. In T. Rothengatter and R. Huguenin (eds), *Traffic and Transport Psychology: Theory and Application,* 147–154. Amsterdam: Elsevier.

Sullman, M. J., Pajo, K. B., and Meadows, M. L. (2003). Factors affecting the risk of crash involvement amongst New Zealand truck drivers. In L. Dorn (ed.) *Driver Behaviour and Training,* 161–173. Aldershot: Ashgate. First International Conference on Driver Behaviour and Training. Stratford-upon-Avon 11–12 November, 2003.

Summala, H. (1988). Risk control is not risk adjustment: The zero-risk theory of driver behaviour and its implications. *Ergonomics, 31,* 491–506.

Sundström, A. (2008). Self-assessment of driving skill – a review from a measurement perspective. *Transportation Research Part F, 11,* 1–9.

Szlyk, J. P., Alexander, K. R., Severing, K., and Fishman, G. A. (1992). Assessment of driving performance in patients with retinitis pigmentosa. *Archives of Ophthalmology, 110,* 1709–1713.

Szlyk, J. P., Fishman, G. A., Severing, K., Alexander, K. R and Viana, M. (1993). Evaluation of driving performance in patients with juvenile macular dystropies. *Archives of Ophthalmology, 111,* 207–212.

Szlyk, J. P., Mahler, C. L., Seiple, W., Edward, D. P., and Wilensky, J.T. (2005). Driving performance of glaucoma patients correlates with peripheral visual field loss. *Journal of Glaucoma, 14,* 145–150.

Szlyk, J. P., Pizzimenti, C. E., Fishman, G. A., Kelsch, R., Wetzel, L. C., Kagan, S., and Ho, K. (1995). A comparison of driving in older subjects with and without age-related macular degeneration. *Archives of Ophthalmology, 113,* 1033–1040.

Szlyk, J. P., Seiple, W., and Viana, M. (1995). Relative effects of age and compromised vision on driving performance. *Human Factors, 37,* 430–436.

Szlyk, J. P., Taglia, D. P., Paliga, J., Edward, D. P., and Wilensky, J. T. (2002). Driving performance in patients with mild to moderate glaucomatous clinical vision changes. *Journal of Rehabilitation Research and Development, 39,* 467–495.

Tarawneh, M. S., McCoy, P. T., Bishu, R. R., and Ballard, J. L. (1993). Factors associated with driving performance of older drivers. *Transportation Research Record, 1405,* 64–71.

Taubman-Ben-Ari, O., Mikulincer, M., and Gillath, O. (2004). The multidimensional driving style inventory – scale construct and validation. *Accident Analysis and Prevention, 36,* 323–332.

Taylor, J., Chadwick, D., and Johnson, T. (1995). Accident experience and notification rates in people with recent seizures, epilepsy or undiagnosed episodes of loss of consciousness. *Quarterly Journal of Medicine, 88,* 733–740.

Taylor, J., Chadwick, D., and Johnson, T. (1996). Risk of accidents in drivers with epilepsy. *Journal of Neurology and Neurosurgical Psychiatry, 60,* 621–627.

Taylor, M. C., and Lockwood, C. R. (1990). *Factors Affecting the Accident Liability of Motorcyclists – a Multivariate Analysis of Survey Data.* Research Report 270. Crowthorne: Transport and Road Research Laboratory.

Terhune, K. W. (1983). An evaluation of responsibility analysis for assessing alcohol and drug crash effects. *Accident Analysis and Prevention, 15,* 237–246.

Tillman, W. A., and Hobbs, G. E. (1949). The accident-prone automobile driver: A study of the psychiatric and social background. *American Journal of Psychiatry, 106,* 321–331.

Tirunahari, V. L., Zaidi, S. A., Sharma, R., Skurnick, J., and Ashtyani, H. (2003). Microsleep and sleepiness: A comparison of multiple sleep latency test and scoring of microsleep as a diagnostic test for excessive daytime sleepiness. *Sleep Medicine, 4,* 63–67.

Toomath, J. B., and White, W. T. (1982). New Zealand survey of driver exposure to risk of accidents. *Accident Analysis and Prevention, 14,* 407–411.

Tranter, P., and Warn, J. (2008). Relationships between interest in motor racing and driver attitudes and behaviour amongst mature drivers: An Australian case study. *Accident Analysis and Prevention, 40,* 1683–1689.

Tronsmoen, T. (2008). Associations between self-assessment of driving ability, driver training and crash involvement among young drivers. *Transportation Research Part F, 11,* 334–346.

Tzamalouka, G., Papadakaki, M., and Chliaoutakis, J. E. (2005). Freight transport and non-driving work duties as predictors of falling asleep at the wheel in urban areas of Crete. *Journal of Safety Research, 36,* 75–84.

Ulleberg, P., and Rundmo, T. (2002). Risk-taking attitudes among young drivers: The psychometric qualities and dimensionality of an instrument to measure young drivers' risk-taking attitudes. *Scandinavian Journal of Psychology, 43,* 227–237.

Underwood, G., Chapman, P., Wright, S., and Crundall, D. (1999). Anger while driving. *Transportation Research Part F, 2,* 55–68.

Valuri, G., Stevenson, M., Finch, C., Hamer, P., and Elliott, B. (2005). The validity of a four week self-recall of sports injuries. *Injury Prevention, 11,* 135–137.

Vance, D. E., Roenker, D. L., Cissell, G. M., Edwards, J. D., Wadley, V. G., and Ball, K. K. (2006). Predictors of driving exposure and avoidance in a field study of older drivers from the state of Maryland. *Accident Analysis and Prevention, 36,* 823–831.

Vassallo, S., Smart, D., Sanson, A., Harrison, W., Harris, A., Cockfield, S., and McIntyre, A. (2007). Risky driving among young Australian drivers: Trends, precursors and correlates. *Accident Analysis and Prevention, 39,* 444–458.

Vernon, D. D., Diller, E. M., Cook, L. J., Reading, J. C., Suruda, A. J., and Dean, J. M. (2002). Evaluating the crash and citation rates of Utah drivers licensed with medical conditions 1992–1996. *Accident Analysis and Prevention, 34,* 237–246.

Verschuur, W. L., and Hurts, K. (2008). Modeling safe and unsafe driving behaviour. *Accident Analysis and Prevention, 40,* 644–656.

Versteegh, S. (2004). The accuracy of driver accounts of vehicle accidents. In *Proceedings of Road Safety Research, Policing and Education Conference.*

Verwey, W. B, and Zaidel, D. M. (2000). Predicting drowsiness accidents from personal attributes, eye blinks and ongoing driver behaviour. *Personality and Individual Differences, 28,* 123–142.

Violanti, J. M., and Marshall, J. R. (1996). Cellular phones and traffic accidents: An epidemiological approach. *Accident Analysis and Prevention, 28,* 265–270.

Vogel, R., and Rothengatter, J. A. (1984). *Motieven van Snelheidsgedrag op Autosnelwegen: Een Attitude Onderzoek* (Motives for speeding behaviour on motorways: An attitudinal study). Report VK 84–09. The Netherlands: Traffic Research Centre, University of Groningen.

Wallace, J. C., and Vodanovich, S. J. (2003). Can accidents and industrial mishaps be predicted? Further investigation into the relationship between cognitive failure and reports of accidents. *Journal of Business and Psychology, 17,* 503–514.

Wallén Warner, H., and Åberg, L. (2008). Driver's beliefs about exceeding the speed limits. *Transportation Research Part F, 11,* 376–389.

Waller, P. F., Elliott, M. R., Shope, J. T., Raghunathan, T. E., and Little, R. J. (2001). Changes in young adult offence and crash patterns over time. *Accident Analysis and Prevention, 33,* 117–128.

Walton, D. (1999). Examining the self-enhancement bias: Professional truck drivers' perceptions of speed, safety, skill and consideration. *Transportation Research Part F, 2,* 91–113.

Warner, M., Schenker, N., Heinen, M. A., and Fingerhut, L. A. (2005). The effects of recall on reporting injury and poisoning episodes in the National Health Interview Survey. *Injury Prevention, 11,* 282–287.

Wasielewski, P. (1984). Speed as a measure of driver risk: Observed speed versus driver and vehicle characteristics. *Accident Analysis and Prevention, 16,* 89–102.

Webb, G. R., Bowman, J. A., and Sanson-Fisher, R. W. (1988). Studies of child safety restraint use in motor vehicles: Some methodological considerations. *Accident Analysis and Prevention, 20,* 109–115.

Weber, D. C. (1972). An analysis of the California driver record study in the context of a classical accident model. *Accident Analysis and Prevention, 4,* 109–116.

Weiss, G., Hechtman, L., Perlman, T., Hopkins, J., and Wener, A. (1979). Hyperactives as young adults: A controlled prospective ten-year follow-up of 75 children. *Archives of General Psychiatry, 36,* 675–681.

West, R. (1995). Towards unravelling the confounding of deviant driving, drink driving and traffic accident liability. *Criminal Behaviour and Mental Health, 5,* 452–462.

West, R. (1997). *Accident Script Analysis.* TRL Report 274. Crowthorne: Transport Research Laboratory.

West, R. (1998). *Accident Rates and Behavioural Characteristics of Novice Drivers in the TRL Cohort study.* TRL Report 293. Crowthorne: Transport Research Laboratory.

West, R., Elander, J., and French, D. (1992). *Decision Making, Personality and Driving Style as Correlates of Individual Crash Risk.* TRRL Contractor Report 309. Crowthorne: Transport and Road Research Laboratory.

West, R. J., Elander, J., and French, D. J. (1993). Mild social deviance, Type-A behaviour pattern and decision-making style as predictors of self-reported driving style and traffic accident risk. *British Journal of Psychology, 84,* 207–219.

West, R. J., French, D. J., Kemp, R., and Elander, J. (1993). Direct observation of driving, self reports of driver behaviour, and accident involvement. *Ergonomics, 36,* 557–567.

West, R., and Hall, J. (1997). The role of personality and attitudes in traffic accident risk. *Applied Psychology: An International Review, 46,* 253–264.

West, R., and Hall, J. (1998). *Accident Liability of Novice Drivers.* TRL Report 295. Crowthorne: Transport Research Laboratory.

White, S. B. (1976). On the use of annual vehicle miles of travel estimates from vehicle owners. *Accident Analysis and Prevention, 8,* 257–261.

Whitlock, G. H., Clouse, R. J., and Spencer, W. F. (1963). Predicting accident proneness. *Personnel Psychology, 16,* 35–44.

Wickens, C. M., Toplak, M. E., and Wiesenthal, D. L. (2008). Cognitive failures as predictors of driving errors, lapses, and violations. *Accident Analysis and Prevention, 40,* 1223–1233.

Wiggins, C. L., Schmidt-Nowara, W. W., Coultas, D. B., and Samet, J. M. (1990). Comparison of self and spouse reports of snoring and other symptoms associated with sleep-apnea syndrome. *Sleep, 13,* 245–252.

Wilde, G. J. (1982). Critical issues in risk homeostasis theory. *Risk Analysis, 2,* 249–258.

Wilde, G. J. (1985). Assumptions necessary and unnecessary to risk homeostasis. *Ergonomics, 28,* 1531–1538.

Wilde, G. J. (1988). Risk homeostasis theory and traffic accidents: Propositions, deductions and discussion of dissension in recent reactions. *Ergonomics, 31,* 441–468.

Willemsen, J., Dula, C. S., Declerq, F., and Verhaeghe, P. (2008). The Dula Dangerous Driving Index: An investigation of reliability and validity across cultures. *Accident Analysis and Prevention, 40,* 798–806.

Williams, A. F., and Shabanova, V. I. (2003). Responsibility of drivers, by age and gender, for motor-vehicle crash deaths. *Journal of Safety Research, 34,* 527–531.

Williams, C. L., Henderson, A. S., and Mills, J. M. (1974). An epidemiological study of serious traffic offenders. *Social Psychiatry, 9,* 99–109.

Williams, M. J. (1981). Validity of the traffic conflicts technique. *Accident Analysis and Prevention, 13,* 133–145.

Williford, W. O., and Murdock, G. R. (1978). Statistical estimation of individual motor vehicle driver's accident liability based on an analysis of the mean time intervals between accidents of Georgia drivers. *Accident Analysis and Prevention, 10,* 189–205.

Wilson, R. J., and Jonah, B. A. (1988). The application of problem behavior theory to the understanding of risky driving. *Alcohol, Drugs and Driving, 4,* 173–191.

Wilson, R. J., Meckle, W., Wiggins, S., and Cooper, P. J. (2006). Young driver risk in relation to parents' retrospective driving record. *Journal of Safety Research, 37,* 325–332.

Wolfe, A. C. (1982). The concept of exposure to the risk of a road traffic accident and an overview of exposure data collection methods. *Accident Analysis and Prevention, 14,* 337–340.

Wong, W. A., and Hobbs, G. E. (1949). Personal factors in industrial accidents – a study of accident proneness in an industrial group. *Industrial Medicine, 18,* 291–294.

Wu, H., and Yan-Go, F. (1996). Self-reported automobile accidents involving patients with obstructive sleep apnea. *Neurology, 46,* 1254–1257.

af Wåhlberg, A. E. (2000). The relation of acceleration force to traffic accident frequency: A pilot study. *Transportation Research Part F, 3,* 29–38.

af Wåhlberg, A. E. (2001). The theoretical features of some current approaches to risk perception. *Journal of Risk Research, 4,* 237–250.

af Wåhlberg, A. E. (2002a). On the validity of self-reported traffic accident data. E140 Proceedings of Soric'02. Available at www.psyk.uu.se/hemsidor/busdriver.

af Wåhlberg, A. E. (2002b). Characteristics of low speed accidents with buses in public transport. *Accident Analysis and Prevention, 34,* 637–647.

af Wåhlberg, A. E. (2003a). Some methodological deficiencies in studies on traffic accident predictors. *Accident Analysis and Prevention, 35,* 473–486.

af Wåhlberg, A. E. (2003b). Stability and correlates of driver acceleration behaviour. In L. Dorn (ed.) *Driver Behaviour and Training*, 45–54. Aldershot: Ashgate. First International Conference on Driver Behaviour and Training. Stratford-upon-Avon 11–12 November, 2003.

af Wåhlberg, A. E. (2004a). Characteristics of low speed accidents with buses in public transport. Part II. *Accident Analysis and Prevention, 36,* 63–71.

af Wåhlberg, A. E. (2004b). The stability of driver acceleration behavior, and a replication of its relation to bus accidents. *Accident Analysis and Prevention, 36,* 83–92.

af Wåhlberg, A. E. (2006a). Speed choice versus celeration behavior as traffic accident predictor. *Journal of Safety Research, 37,* 43–51.

af Wåhlberg, A. E. (2006b). Driver celeration behavior and the prediction of traffic accidents. *International Journal of Occupational Safety and Ergonomics, 12,* 281–296.

af Wåhlberg, A. E. (2007a). Effects of passengers on bus driver celeration behavior and incident prediction. *Journal of Safety Research, 38,* 9–15.

af Wåhlberg, A. E. (2007b). Aggregation of driver celeration behavior data: Effects on stability and accident prediction. *Safety Science, 45,* 487–500.

af Wåhlberg, A. E. (2007c). Making bad look good: The scientific claims of Interactive Driving Systems. Available at www.psyk.uu.se/hemsidor/busdriver.

af Wåhlberg, A. E. (2007d). Long-term prediction of traffic accident record from bus driver celeration behavior. *International Journal of Occupational Safety and Ergonomics, 13,* 159–171.

af Wåhlberg, A. E. (2008a). The relation of non-culpable traffic incidents to bus drivers' celeration behavior. *Journal of Safety Research, 39,* 41–46.

af Wåhlberg, A. E. (2008b). Driver celeration behavior and accidents – an analysis. *Theoretical Issues in Ergonomics Science, 9,* 383–403.

af Wåhlberg, A. E. (2009). Hourly changes in accident risk for bus drivers. *Journal of Risk Research, 12,* 187–197.

af Wåhlberg, A. E. (submitted). Social desirability effects in driver behaviour inventories.

af Wåhlberg, A. E., and Dorn, L. (2007). Culpable versus non-culpable traffic accidents; what is wrong with this picture? *Journal of Safety Research, 38,* 453–459.

af Wåhlberg, A. E., and Dorn, L. (2009a). Bus driver accident record; the return of accident proneness. *Theoretical Issues in Ergonomics Science, 10,* 77–91.

af Wåhlberg, A. E., and Dorn, L. (2009b). Absence behaviour as traffic crash predictor in bus drivers. *Journal of Safety Research,40,* 197–201.

af Wåhlberg, A. E., and Dorn, L. (submitted). Bus driver accident record; stability over time, exposure and culpability.

af Wåhlberg, A. E., Dorn, L., and Kline, T. (forthcoming). The Manchester Driver Behaviour Questionnaire as traffic accident predictor. *Theoretical Issues in Ergonomics Science.*

af Wåhlberg, A. E., Dorn, L., and Kline, T. (submitted). The effect of social desirability on self reported and recorded road traffic accidents; Implications for self reports of driver behaviour.

af Wåhlberg, A. E., and Melin, L. (2008). Driver celeration behavior in training and regular driving. In L. Dorn (ed.) *Driver Behaviour and Training, Volume III,* 189–199. Aldershot: Ashgate. Third International Conference on Driver Behaviour and Training. Dublin 12–13 November, 2007.

Ysander, L. (1966). The safety of drivers with cronic disease. *British Journal of Industrial Medicine, 23,* 28–36.

Zhang, J., Lindsay, J., Clarke, K., Robbins, G., and Mao, Y. (2000). Factors effecting the severity of motor vehicle traffic crashes involving elderly drivers in Ontario. *Accident Analysis and Prevention, 32,* 117–125.

Zhao, J., Mann, R. E., Chipman, M., Adlaf, E., Stoduto, G., and Smart, R. G. (2006). The impact of driver education on self-reported collisions among young drivers with a graduated license. *Accident Analysis and Prevention, 38,* 35–42.

Zimbardo, P. G., Keough, K. A., and Boyd, J. N. (1998). Present time perspective as a predictor of risky driving. *Personality and Individual Differences, 23,* 1007–1023.

van Zomeren, A., Brouwer, W., Rothengatter, J. A., and Snoek, J. W. (1988). Fitness to drive a car after recovery from severe head injury. *Archives of Physical Medicine and Rehabilitation, 69,* 90–96.

Zwerling, C., Sprince, N. L., Wallace, R. B., Davis, C. D., Whitten, P. S., and Heeringa, S. G. (1995). Effect of recall period on the reporting occupational injuries among older workers in the Health and Retirement Study. *American Journal of Industrial Medicine, 28,* 583–590.

Zylman, R. (1972). Drivers' records: Are they a valid measure of driving behaviour? *Accident Analysis and Prevention, 4,* 333–349.

Zylman, R. (1974). A critical evaluation of the literature on 'alcohol involvement' in highway deaths. *Accident Analysis and Prevention, 6,* 163–204.

Åberg, L., Larsen, L., Glad, A., and Beilinsson, L. (1997). Observed vehicle speed and drivers' perceived speed of others. *Applied Psychology: An International Review, 46,* 287–302.

Åberg, L., and Rimmö, P.-A. (1998). Dimensions of aberrant driver behaviour. *Ergonomics, 41,* 39–56.

Åberg, L., and Wallén Warner, H. (2008). Speeding – deliberate violation or involuntary mistake? *Revue Européenne de Psychologie Appliqée, 58,* 23–30.

Index